Systems Biology and Machine Learning Methods in Reproductive Health

Systems Biology and Machine Learning Methods in Reproductive Health is an innovative and wide-ranging book that discovers the synergetic combination of disciplines: systems biology and machine learning, with an application in the field of reproductive health. This book assembles the expertise of leading scientists and clinicians to present a compilation of cutting-edge techniques and case studies utilizing computational methods to elucidate intricate biological systems, elucidate reproductive pathways, and address critical issues in the fields of fertility, pregnancy, and reproductive disorders.

Bringing science and data science together, this groundbreaking book provides scientists, clinicians, and students with a step-by-step guide to uncovering the complexities of reproductive health through cutting-edge computational tools.

Computational Biology Series

About the Series:

This series aims to capture new developments in computational biology, as well as high-quality work summarizing or contributing to more established topics. Publishing a broad range of reference works, textbooks, and handbooks, the series is designed to appeal to students, researchers, and professionals in all areas of computational biology, including genomics, proteomics, and cancer computational biology, as well as interdisciplinary researchers involved in associated fields, such as bioinformatics and systems biology.

Computational Genomics with R
Altuna Akalin, Bora Uyar, Vedran Franke and Jonathan Ronen

An Introduction to Computational Systems Biology: Systems-level Modelling of Cellular Networks
Karthik Raman

Virus Bioinformatics
Dmitrij Frishman and Manuela Marz

Multivariate Data Integration Using R: Methods and Applications with the mixOmics Package
Kim-Anh LeCao and Zoe Marie Welham

Bioinformatics: A Practical Guide to NCBI Databases and Sequence Alignments
Hamid D. Ismail

Data Integration, Manipulation and Visualization of Phylogenetic Trees
Guangchuang Yu

Bioinformatics Methods: From Omics to Next Generation Sequencing
Shili Lin, Denise Scholtens, and Sujay Datta

Systems Medicine: Physiological Circuits and the Dynamics of Disease
Uri Alon

The Applied Genomic Epidemiology Handbook: A Practical Guide to Leveraging Pathogen Genomic Data in Public Health
Allison Black and Gytis Dudas

For more information about this series please visit: www.routledge.com/Chapman--Hall CRC-Computational-Biology-Series/book-series/CRCCBS

Systems Biology and Machine Learning Methods in Reproductive Health

Edited by

Abhishek Sengupta, Priyanka Narad,
Dinesh Gupta, and Deepak Modi

CRC Press is an imprint of the
Taylor & Francis Group, an **informa** business
A CHAPMAN & HALL BOOK

First edition published 2025
by CRC Press
2385 NW Executive Center Drive, Suite 320, Boca Raton, FL 33431

and by CRC Press
4 Park Square, Milton Park, Abingdon, Oxon, OX14 4RN

CRC Press is an imprint of Taylor & Francis Group, LLC

ISBN: 9781032755519 (hbk)
ISBN: 9781032783703 (pbk)
ISBN: 9781003487548 (ebk)

DOI: 10.1201/9781003487548

Typeset in Minion
by Newgen Publishing UK

Contents

Preface

Systems Biology and Machine Learning Methods in Reproductive Health is a comprehensive book that explores the application of cutting-edge computational approaches to address challenges in reproductive health. It provides an introduction to systems biology and machine learning principles and explores various data sources and integration methods specific to reproductive health research. The book covers genomics, transcriptomics, proteomics, and metabolomics studies in this domain, highlighting systems-level analyses. It discusses machine learning algorithms tailored for reproductive health applications, paving the way for personalized medicine approaches. The book also examines ethical and privacy considerations surrounding the use of these technologies in healthcare. Finally, it outlines key challenges and offers insights into future directions for harnessing systems biology and machine learning to advance reproductive health research and clinical practice.

This book aims to provide readers with an extensive understanding of data-driven methodologies applied to reproductive health research. Exploring systems biology and machine learning techniques offers insights into leveraging complex data for innovative solutions in this domain. The authors anticipate that readers will find the content of this book informative and beneficial. Furthermore, we welcome constructive feedback and suggestions from readers, as their perspectives can enrich the discourse surrounding the topics covered in this volume.

Abhishek Sengupta
Noida, Uttar Pradesh, India

Priyanka Narad
New Delhi, India

Dinesh Gupta
New Delhi, India

Deepak Modi
Mumbai, India

Editors' Biographies

Abhishek Sengupta is Assistant Professor at the Centre of Computational Biology and Bioinformatics, Amity Institute of Biotechnology, Amity University, Noida, Uttar Pradesh, India. He received his MSc from Nottingham Trent University, UK, and PhD from Amity University, India. With 15 years of experience in genome-scale metabolic reconstructions, constraint-based modeling, network biology, systems biology, metabolomics, and flux balance analysis, he has developed the HEPNet knowledge base, contributed to software like TFIS, ARTPre, and PluriMetNet, and databases like VIRdb. He has published extensively in reputed journals and received grants from DST-SERB (a statutory body of the Department of Science and Technology), Government of India. Sengupta has also led technology transfer and licensing of the ML-based software "FertilitY Predictor" and been awarded copyrights for "ARTPre: An Online Tool to Predict the Success Rates of Assisted Reproductive Procedures in Indian Subcontinent".

Priyanka Narad is an experienced Bioinformatician and AI expert. She is currently working as a Scientist at the Indian Council of Medical Research (ICMR), New Delhi. With over 13 years of experience, including as an assistant professor at Amity University, her expertise lies in stem cell bioinformatics, machine learning, multi-omics data integration, and predictive modeling. Narad has made significant contributions through numerous publications in prestigious journals like *Nature Scientific Reports* and *PeerJ*. She has secured funding for projects like "A Hybrid Bayesian Approach to Address Socio-Economic Challenges in Assisted Reproductive Techniques Across the Indian Sub-population." Narad has developed and deployed software/databases such as TFIS, ARTPre, VIRdb, and FertilitY Predictor, for which she holds technology transfer and copyright licenses. Her academic excellence was recognized with the DST SERB Young Scientists Travel Award to attend a systems biology course at EMBL-EBI (European Molecular Biology Laboratory-European Bioinformatics Institute), UK.

Dinesh Gupta is a distinguished bioinformatician and computational biologist. He obtained his PhD from All India Institute of Medical Sciences in New Delhi. He currently serves as Group Leader of the Translational Bioinformatics Group at the International Centre for Genetic Engineering and Biotechnology (ICGEB) in New Delhi, India. With over two decades of experience in the field, Gupta has made significant contributions to the development and application of bioinformatics tools and artificial intelligence methods for solving complex biological problems. His research interests span a wide range of areas, including machine learning for biological data analysis, computer-aided drug design, comparative genomics, systems biology, and next-generation sequencing data analysis. Gupta has published extensively in prestigious journals and has been actively involved in organizing international bioinformatics workshops and training programs.

Deepak Modi is a renowned scientist in Reproductive Biology and Genetics, currently associated with the National Institute for Research in Reproductive and Child Health, Indian Council of Medical Research (ICMR). With a PhD from the University of Mumbai and an extensive academic background, he has received prestigious awards like the PM Bhargava Oration Award and GP Talwar Middle Career Scientist Award. His research focuses on embryo implantation, infertility, and disorders of sex development. Modi has an impressive publication record with 85 publications and 3 book chapters in reputed journals. He has contributed significantly through projects on topics like immunomodulatory roles of HOXA10, microfluidic placental function assessment, endometriosis pathogenesis, and COVID-19 placenta. In addition, he actively participates in scientific conferences and serves as an invited speaker and panelist, reflecting his expertise in the field.

CHAPTER 1

Introduction to Systems Biology and Machine Learning

Vandana C. D., Krupa S., and Sandeep Kumar C.

1.1 INTRODUCTION

The Human Genome Project (HGP) brought a revolution in medicine and has fundamentally altered the field of biology. The concept underlying the HGP first came forth from Renato Dulbecco in a 1986 paper (Dulbecco 1986), where he pointed out that understanding the human genome's sequence would improve our understanding of cancer (Birney 2021).

The HGP's most significant influence is that it has inspired scientists to embrace an alternative viewpoint on biology referred to as the systems approach. Instead of studying individual genes or proteins one at a time, systems biology has allowed us to understand the big picture of the cell's functioning. System biology has achieved this by promoting a fresh scientific methodology in biology known as discovery science; outlining a catalogue of genetic components for humans and various model organisms; reinforcing the notion that biology operates as an informational science; by equipping us with advanced high-throughput tools for methodically disturbing and observing biological systems; and by inspiring the development of novel computational techniques (Birney 2021).

The goal of the emerging subject of systems biology, is to comprehend biological systems at the system level. A collection of guiding ideas and techniques that connect molecular behaviors to the features and operations of systems is necessary for system-level comprehension. In the end, a coherent body of knowledge supported by fundamental physics will be used to characterize and comprehend cells, organisms, and humans at the system level (Kitano 2002).

DOI: 10.1201/9781003487548-1

1.1.1 History

This approach toward biology has been pursued for a long time and this concept has emerged several times in the biological scientific community. The earliest and foremost proponents of system-level understanding of biology were credited to Norbert Wiener. He was an American scientist who worked in the fields of applied mathematics and also the inverter of cybernetics or biological cybernetics branch of science. In 1968, Ludwig Von Bertalanffy developed the concept of general system theory that helped to understand the systems. However, this theory was not well developed and too abstract to be accepted by a scientific audience. The basic form of this concept was proposed by Cannon and he also proposed the concept of homeostasis (Kitano 2002). During the above timeline, the field of molecular biology did not provide them with enough knowledge to understand this phenomenon. Therefore, the entire work was focused on explaining them at the physiological level. Systems biology is different from previous attempts, in that it offers the chance to directly base system-level understanding on molecular-level information, such as genes and proteins. Previous attempts have not been able to adequately connect molecular-level knowledge to system-level description. This presents a unique chance to comprehend biological systems through a coherent framework of knowledge that has been accumulated from the molecular to the system level (Kitano 2002).

1.1.2 Scope

The reach of systems biology in basic research is very broad, which includes multiple targets. Different sets of tools and techniques are required for each research target. Furthermore, it also needs information from different areas of biology such as engineering fields, computer theory, control theory, high-precision measurement molecular biology, and other scientific areas. Research is mainly carried out in four areas (Figure 1.1).

1.1.3 Key Properties

To obtain systems-level insight into biological properties, knowledge needs to be gained from four different areas:

1.1.3.1 Systems Structure

The biological systems are composed of multiple components: genes, lipids, carbohydrates, and proteins. These molecules are involved in intramolecular interactions and intermolecular interactions. Furthermore, some of the biomolecules can also form a network that is essential for cell functioning. These networks significantly change the overall behavior of these molecules from their monomeric form (Kitano 2002).

1.1.3.2 System Dynamics

All systems are subjected to various conditions during their lifetime and are associated with respective metabolic changes in internal organelles. These internal changes are measured by phase portrait and bifurcation analysis, which uses techniques like sensitivity analysis, dynamic analysis, and metabolic analysis. These measurements allow us to

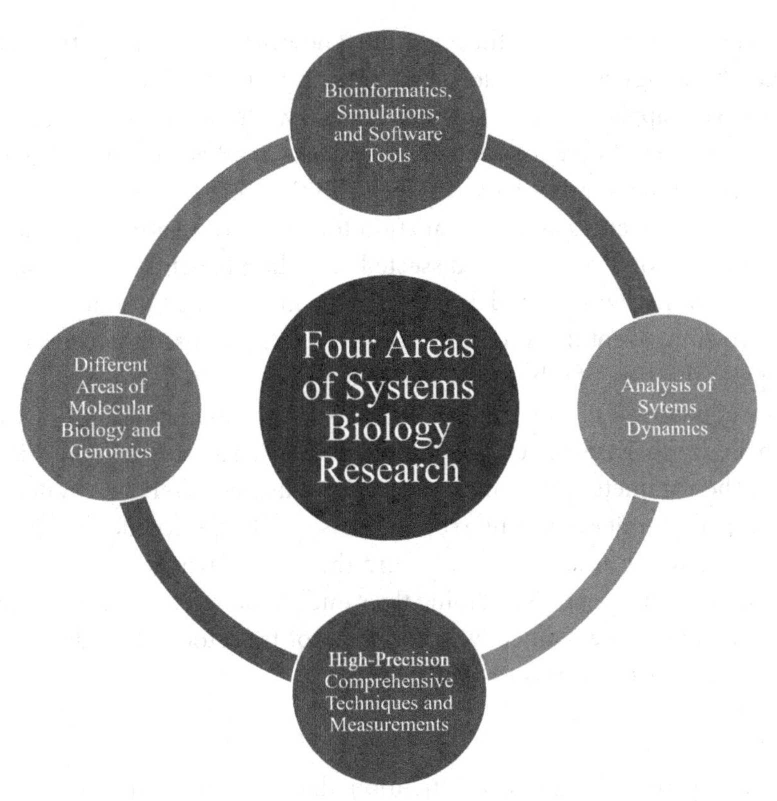

FIGURE 1.1 Four major areas of systems biology.

identify specific molecular mechanisms underpinning specific cell behavior. In multidimensional space, bifurcation analysis is used to understand time-related changes in the biological system. Furthermore, each dimension is correlated with the concentration of a particular biochemical metabolite in the cell (Kitano 2002).

1.1.3.3 Control Method

The cell's state is controlled by several mechanisms. These mechanisms can be made robust to either reduce or prevent alterations that are caused by the malfunctioning, intrinsic, and extrinsic factors that are responsible for this malfunctioning. Further, this also highlights potential therapeutic areas for a particular disease (Kitano 2002).

1.1.3.4 Design Method

It deals with innovative strategies to alter and design the biological system with task-specific properties. Instead of traditional trial and error methods, this approach depends on a well-defined framework. The fundamental concepts are design principles and simulations.

The advancement in the above four areas requires needs innovations in genomics, computational sciences, and measurement techniques, and combining these innovations with existing and new knowledge is the way forward in genomics (Kitano 2002).

Systems biology tries to answer the question of how molecular interactions are responsible for the physiology and behavior of an organism. In previous centuries, mathematical models were applied in genetics, biochemistry, physiology, evolutionary biology, and ecology. However, different scientists—physicists, mathematicians, engineers, computer scientists, and biologists—have diverse definitions of models. If model building is to reach its full effectiveness as an explanation tool for investigating biological systems, these distinct points of view must be dissected, and their benefits must be summarized. The complexity of mechanical models is due to the involvement of numerous unknown parameters. To circumvent this problem, scientists have proposed the concept of robustness in the past. This idea of robustness was widely accepted in the scientific community; however, the application of this concept still lacks a precise definition (Tavassoly et al. 2018). A challenge that has arisen in systems biology is robustness, and it is especially applicable to the parameter problem. In general, robustness refers to a system's capacity to preserve a specific trait despite interruption. It is vital to specify the nature of the property, the sense in which it is unchanging, and the kind of perturbations that are being considered to make this clear. "Remaining the same" could refer to the quantity and kind of attractors as well as the connectivity and form of trajectories that do not change in response to disturbances (Tavassoly et al. 2018).

1.1.4 Types of Robustness

Robustness has garnered significant attention due in part to the possibility that it represents a biological design concept (Kitano 2002). As demonstrated, various forms of resilience may be categorized based on the dynamical system's altered features (Table 1.1).

1.1.5 Complex System

Both artificial and natural complex systems display universal design patterns, which are shared ideas and guidelines that show up in several, seemingly unrelated systems. The successful construction of complex systems that can compete, live, reproduce, and evolve over long periods of time and several generations—thereby promoting enhanced fitness and overall growth—requires the use of these design patterns. Understanding these recurrent design principles—which appear in a variety of complex systems and contexts, both natural and man-made—is the goal of the area of complex systems theory.

TABLE 1.1 Types of resilience.

Various Forms of Resilience	Functions
Type I stability in motion	Ability to withstand starting conditions while maintaining consistent constrain values.
Type II resilience to constrain change	Resistance against alterations of constrained values
Type III resilience to constrain change	Robustness against modulation in parametric values
Type IV stability in structure	Resilience to alteration in dynamic functioning

Accurately describing the characteristics of complex systems is the aim of complex systems science, to provide a complete understanding that goes beyond individual systems or particular design concepts. Gaining a deeper comprehension of these universal principles will help us better navigate and understand the rapid changes brought about by the social and technical growth in our environment (Ma'ayan 2017). Scientists try to analyze and grasp complicated systems by conducting multivariate experiments. These studies capture measurements of numerous system variables under controlled settings to track the dynamics of the system over time in response to various perturbations. Consequently, the gathered data is used to construct models, which are necessary to produce hypotheses that make sense of the data that has been seen. These models provide a simplified representation, or skeleton, of the real complicated system that is being studied to gain insights into cell function (Ma'ayan 2017). The goal of modeling in the study of complex systems is to reduce the real system's complexity to a manageable scale that can be theoretically, mathematically, and cognitively understood.

1.1.6 Environmental Complexity versus Agents' Complexity

Complex environments and complex agents are the two basic types that can be identified in the context of complex systems and discussions on complex system design. Systems having well-defined borders, surrounded by a physical border, are referred to as complex agents. These agents often have one or more CPUs, a clock, energy-efficient acquisition and use mechanisms, sensors, actuators, and environmental interaction capabilities. Complex agents can move, grow, heal themselves, reproduce themselves, and frequently even be conscious of their own existence. Humans, cells, trees, fish, worms, birds, cars, airplanes, and some types of robots are examples of complex agents (Ma'ayan 2017).

1.1.7 From Molecules to Pathways and Networks

Large-scale molecular datasets are produced by experimental omics research, and statistical techniques are necessary to pinpoint the genes, messenger RNA (mRNA), proteins, and other molecular elements that are connected to particular physiological or pathological conditions. Based on their under- or over-representation in biological samples, Gene Set Enrichment Analysis (GSEA) is a statistical method used to find putative molecular components accountable for phenotypes and functions (Ashburner et al. 2000).

A particular ontology is used to enrich differentially expressed molecular entities, or more generally, biomarkers. An ontology is a collection of concepts that are organized into certain relationships and function as a hierarchical classifier. By linking data (genes or gene products) with biological processes, molecular functions, and cellular components, this ontology serves as a tool for the discovery of biological knowledge (Ashburner et al. 2000).

Systems biology uses a variety of ontologies, including Molecular Biology of the Cell Ontology (MBCO) and Gene Ontology (GO). Furthermore, data is converted into biological information through the use of bioinformatics tools such as the Kyoto Encyclopedia of Genes and Genomes (KEGG), WikiPathways, Reactome Pathway, Progeny Signatures, and Broad Signatures. The GSEA results offer valuable information about the pathways, such as signaling pathways, that govern the particular phenotype being studied. This method makes it easier to identify important molecular actors and their roles in the biological processes connected to the phenotype under study (Ashburner et al. 2000).

Within cells, signaling pathways are essential for information processing. These pathways take in signals from the outside world and use them to control the physiology of cells. Signaling pathways are complex networks made up of multiple components that can all receive, transmit, and transduce information (Tavassoly et al. 2018). Cellular signals carry information that moves through space and time. Differential equations and dynamical systems theory can be used to study this process mathematically. Connectivity between different routes within a network is established by receptors, which are found on the surfaces of cells and are in charge of receiving signals from the external environment as well as other internal signaling components. These intracellular signaling components serve as effectors, integrators of signals, and information processing units within the cell.

They are output devices that show how cells react to signals from outside the cell. Understanding the kinetics of signaling pathways contributes to our comprehension of how cells perceive and react in unison to environmental stimuli. GSEA enables the design and analysis of functional molecular networks in addition to linear routes. The building blocks of these networks are molecular entities interacting with one another; these entities are called "nodes," and the interactions are called "edges." These networks of interactions include a variety of linkages, including direct binding that triggers the activation or inhibition of enzymatic activity and downstream targets. Beyond linear pathways, the network architecture offers a more thorough understanding of the links and linkages between molecular components, revealing details about the intricacy of cellular functions (Hansen et al. 2017).

1.1.8 Machine Learning and Systems Biology

It is a difficult undertaking to evaluate the druggability of all proteins or nucleic acid fragments through experimentation. The problem is made more difficult by our ignorance of the molecular biology of disease. We have a very large sample space for a possible therapeutic target with these uncertainties. Therefore, it is not feasible to clinically examine every pharmacological target before having the ability to rank them in order of importance. Owing to these characteristics, it would be greatly appreciated if computational models could predict therapeutic targets on a genome-wide scale with high sensitivity and high specificity (Kandoi et al. 2015).

We now have access to a vast amount of data, such as profiles of protein and gene expression, metabolic and gene regulatory networks, protein–protein interaction (PPI), and other system-level data, because of technological improvements. While integrating these disparate data sources remains difficult, advancements have been achieved in recent years. These system-level data may now be combined with machine learning and other data mining methods to create predictive models. According to Costa et al. (2010), these studies may reveal biologically significant patterns that grant draggability to prospective therapeutic targets.

A machine learning (ML) method typically involves the subsequent procedures to develop predictive models: choosing memory-based learning (e.g., non-druggable and druggable protein molecules) and ascribing for example: system-level landscapes); choosing memory-based algorithms; and assessing the predictive performance of the systems' mathematical models. The following steps are how we organized this mini-review: The learning cases are covered first, followed by the characteristics and performance measures associated with the system-level-based prediction of draggability, and lastly, the most popular ML techniques in this domain (Kandoi et al. 2015).

Sequence and structural features are among the many characteristics that have been employed to create models that can forecast draggability. Kandoi et al.'s (2015) study paid attention to attributes at the system level, such as the gene expression profile and network topological aspects. To compute the topological properties of networks (henceforth referred to as network measures) to be used as learning features in an ML technique, the PPI networks from which these measures are formed must first be constructed (Kandoi et al. 2015).

Drug-targeting proteins exhibit certain measures of their PPI network that are notably different from the rest of the proteins in the PPI network according to Yao and Rzhetsky (2008) and Jeon et al. (2014). These findings are based on the work of these pioneers in the field of drug-target networks. Therefore, in ML methodologies, network metrics have been demonstrated to be promising predictors of draggability as well as essential and disease genes. ML techniques based on various combinations of PPI topological properties appear to be encouraging for drug-able protein and gene prediction.

The area under the receiver operating characteristic curve (AUC) of prediction models built only using network measurements was 69.21% and about 68%, as shown by Jeon et al. (2014), respectively. Jeon et al. (2014) improved the AUC to 78% by integrating genomic parameters such as mutation occurrence, row chromosomal copy number, Robust Multi-array Average (RMA) intensity, and Genetic Algorithm for Rule-set Prediction (GARP) score with closeness centrality. From this data, it may be inferred that drug-sensitive proteins can be identified only somewhat from PPI measures alone. To quantify the degree of likelihood of drug-sensitive proteins using PPI measures, thorough research will be needed, in which network measurements are employed both singly and collectively as learning qualities.

Yao and Rzhetsky (2008) employed various ML techniques to attain AUCs of 60–72%, taking into account gene expression profile, connectedness, and betweenness, along with other aspects of the system. Costa et al. (2010) attained a median AUC of 82% with their prediction models, which were based on gene expression profile, subcellular localization, and different network measures. These models correctly recovered 78.2% of the known targets at 74.8% precision. They discovered that proteins that are centrally placed within a network of transcription regulators had a higher likelihood of being a drug target after analyzing the characteristics crucial to differentiating druggable from non-druggable genes. By determining the betweenness centrality of each gene in the transcriptional regulatory network, the centrally placed genes were identified. Even though each study used a distinct set of network topological parameters, it appears that pharmacological targets have stronger central connections and are more centralized than the typical gene. Thus, they are the best targets for future work (Costa et al. 2010).

1.2 SYSTEMS BIOLOGY AND ITS RELEVANCE IN REPRODUCTIVE HEALTH: CASE STUDIES

Systems biology takes a more holistic approach, considering the complex interactions and networks that determine how biological systems function. This is particularly valuable in understanding the intricate and interconnected world of reproductive health. Systems biology integrates various—omics data, including genomics, proteomics, metabolomics, and transcriptomics, to create a comprehensive understanding of biological systems (Zhao et al. 2020). It utilizes computational modeling and analysis to unravel the complex relationships between different components and predict how the system might behave under different conditions. Reproductive health encompasses a wide range of aspects, from fertility and pregnancy to sexual dysfunction and age-related decline in reproductive function. Systems biology offers a powerful tool to tackle these challenges by understanding the root causes of infertility. By analyzing the interplay of genes, hormones, and environmental factors, systems biology can help identify the underlying mechanisms behind infertility in both men and women. By developing personalized treatment strategies by creating patient-specific models, systems biology can predict how individuals might respond to different fertility treatments, paving the way for personalized medicine in reproductive healthcare. By improving assisted reproductive technologies (ARTs), systems biology can optimize ART procedures by providing insights into embryo development, implantation, and pregnancy success. Understanding pregnancy complications by modeling the complex interactions between maternal and fetal systems, systems biology can help predict and prevent pregnancy complications like preeclampsia and gestational diabetes. The systems biology approach identified several genes involved in immune function and inflammation that contribute to the development of endometriosis. By predicting the success of in vitro fertilization (IVF) by analyzing gene expression patterns in embryos, researchers can develop models to predict their viability and potential for successful implantation. Understanding the impact of environmental toxins on reproductive health systems biology can be used to study how exposure to environmental pollutants like bisphenol A (BPA) disrupts hormonal signaling and affects fertility (Cantonwine et al. 2013).

1.2.1 Systems Biology Approaches for Infertility and Endometriosis

Infertility diagnosis and treatment have focused on identifying and addressing individual factors, such as hormonal imbalances, anatomical abnormalities, or genetic mutations. However, this approach often overlooks the intricate interplay between these factors and fails to capture the dynamic nature of the reproductive system. Systems biology takes a comprehensive view, considering the complex network of interactions between genes, proteins, metabolites, and signaling pathways that ultimately influence fertility. This approach utilizes a variety of high-throughput technologies, such as Genomics, which is the sequencing of the entire genome to identify genetic variations associated with infertility. Transcriptomics deals with the analysis of gene expression patterns to understand how genes are turned on or off in different cell types and under different conditions. Proteomics is the study of the entire set of proteins expressed in a cell or tissue to understand their functions and interaction and Metabolomics is used to measure the levels of small molecules, such as metabolites, to gain insights into cellular metabolism and energy production (Ding and Schimenti 2021).

By integrating data from these diverse omics platforms, systems biology allows researchers to develop a more complete understanding of the underlying causes of infertility, identify novel diagnostic markers and therapeutic targets, personalize treatment plans based on an individual's unique genetic and molecular profile, and predict the likelihood of successful conception with different ARTs. Researchers are using systems biology approaches to understand the molecular mechanisms contributing to the development and progression of endometriosis, a major cause of female infertility. Systems biology is helping to unravel the complex hormonal and metabolic imbalances associated with polycystic ovarian syndrome (PCOS), which can affect ovulation and egg quality. Studying the sperm proteome and metabolome can provide insights into factors that influence sperm motility, morphology, and DNA integrity, all of which are crucial for successful fertilization. Systems biology is being used to identify molecular markers that predict embryo viability and implantation potential, improving the success rates of IVF and other ARTs.

Researchers are identifying gene signatures and metabolic pathways associated with a receptive endometrium, the lining of the uterus where the embryo implants. This could lead to tests to predict implantation success and guide embryo transfer timing. Systems biology approaches are being used to develop non-invasive methods for assessing embryo quality beyond traditional morphology grading. This could improve embryo selection and reduce reliance on potentially stressful transfer procedures. Recent studies investigate the role of epigenetic modifications in embryo development and implantation. Understanding these modifications could lead to therapies for improving embryo viability and reducing IVF complications. Studies have identified pathways involved in the differentiation of mesenchymal stem cells into endometriotic cells, potentially leading to new therapies targeting this process. Systems biology approaches are investigating the complex interplay between immune cells and inflammatory mediators in endometriosis, offering therapeutic targets for reducing pain and inflammation. Recent research explores the role of epigenetic modifications in the development and persistence of endometriosis, suggesting potential avenues for epigenetic therapy.

1.2.2 Systems Biology Approaches for Preeclampsia

Preeclampsia, a life-threatening pregnancy complication characterized by high blood pressure and proteinuria after 20 weeks of gestation, poses significant challenges for both mothers and babies. Traditionally, studying preeclampsia involved focusing on individual genes or proteins, often yielding limited results due to its multifactorial nature. Systems biology holds immense promise for unraveling the complexities of preeclampsia by analyzing vast datasets of omics data, researchers can pinpoint key pathways and networks implicated in the disease's development. Building computational models that mimic preeclampsia's progression allows simulation of the effects of various genetic and environmental factors, providing valuable insights into the disease mechanisms. Studies revealed four distinct molecular subtypes of preeclampsia, each with unique gene signatures and underlying pathways. This suggests personalized treatment approaches that might be necessary in the future. Another study employed systems biology to identify a network of genes involved in angiogenesis (blood vessel formation). Abnormal angiogenesis is a hallmark of preeclampsia, and this finding sheds light on its potential role in the disease's pathogenesis. Researchers are actively developing computational models of preeclampsia to evaluate the efficacy of different treatment options. These models can personalize care and optimize therapeutic interventions for individual patients (Odenkirk et al. 2020).

High-throughput investigations in preeclampsia can be roughly classified into three types of applications, regardless of the molecular profiling type (global, single-cell, cell-free, etc.) or the type of samples (placenta, blood, etc.) used. We refer to the first application as a class comparison. The purpose of the study is to compare the molecular profiles of patients with clinically defined phenotypes such as all cases of early-onset or late-onset preeclampsia with those of controls. This approach facilitates the deduced pathways and biological processes disrupted in cases linked to the observed symptoms and may lead to the discovery of potential therapeutic targets. Class prediction is the second kind of application. To create disease prediction models, this makes use of ML techniques and discriminant analysis. The goal is not to analyze the changes in molecular profiles that are revealed but rather to maximize prediction accuracy and parsimony. Class discovery is necessary because preeclampsia is a syndromic condition that is characterized by significant variation in expression profiles. This has posed issues for both class comparison and class prediction applications (Chang et al. 2023).

1.2.3 Systems Biology Approaches for Gestational Diabetes

Gestational diabetes mellitus (GDM), affecting 3–8% of pregnant women, poses a significant threat to both mother and child. Any level of glucose intolerance that develops or is identified during pregnancy is known as GDM, one of the most prevalent metabolic disorders of pregnancy. While traditionally studied through individual genes or proteins, its multi-factorial nature demands a holistic approach. Systems biology is a powerful framework offering new insights into GDM's complexities. By integrating omics data, researchers can identify key pathways and regulatory networks involved in insulin resistance and metabolic dysregulation. Given that GDM is a metabolic disorder, metabolomics research is ideally adapted to track changes in tiny molecular metabolites brought on by maternal stimulation or disruption in real time. Metabolomics used for GDM can identify biomarkers for diagnosis, assess the prognosis, direct dietary or medication application, assess the efficacy of treatment, and investigate the process. Building computational models of GDM progression allows for simulating the effects of genetic and environmental factors, providing valuable insights into disease mechanisms and personalized treatments (Wang et al. 2021).

A study identified a network of long non-coding RNAs (lncRNAs) and microRNAs involved in GDM development. This suggests that lncRNAs might be potential therapeutic targets or biomarkers for early diagnosis. Researchers developed a computational model of GDM metabolic pathways. This model helped identify key regulatory nodes and predict the effects of different treatment strategies, paving the way for personalized approaches. Systems biology approaches are being used to understand the role of placental dysfunction in GDM. This could lead to novel interventions targeting the placenta to improve maternal and fetal outcomes (Zhu et al. 2022).

An essential component of the biological system, metabolomics primarily examines blood, urine, and feces before delving into the small-molecule metabolites of diverse metabolic pathway matrices and byproducts. The serum in the amniotic fluid and umbilical cord is also taken for investigation in studies of gestation-associated illnesses like GDM. Mothers' hair or placentas are occasionally taken for examination. Many studies of metabolite profiles have made it easier to identify putative molecular pathways for GDM and diabetes, assisting in the identification of both conditions' parallels and differences. Clinical practice was made possible by Zhu et al.'s investigation of metabolomics indicators and the creation of a panel for the early detection of GDM. A novel perspective on the beginnings and progression of diseases is provided by the developing field of metabolomics. Metabolomics has been used often in different gestational phases in recent GDM investigations. It is expected that metabolomics will play a significant role in the various and early diagnosis of GDM. Pregnancy-related metabolic changes can be dynamically monitored because of longitudinal metabolomics. Furthermore, prognosis and metabolic backgrounds vary amongst GDM patients with distinct physiologic subtypes. Metabolomics would be clinically significant when taking into account clinical indices and physiologic subtypes of GDM.

Carbohydrates, amino acids, lipid metabolites (fatty acids (FAs), phospholipids, sphingomyelin, etc.), purines, uric acid, bile acid, and other small molecular metabolites, as well as associated metabolic pathways, are the core subjects of GDM metabolomics study. The anomalies of small molecule metabolites such as lipids, amino acids, carbohydrates, sterol hormones, bile acids, and disrupted metabolic pathways are the main focus of metabolomics research in GDM. Furthermore, metabolomics research includes the examination of metabolic and signaling pathways based on data from substances that have been found. This helps to better understand the underlying biological phenomena of these pathways by facilitating the investigation of linked metabolites, enzymes, and genes (Spanou et al. 2022).

The various periods of the individuals determine the clinical utility of metabolomics in GDM. The primary clinical utility of research focusing on the first trimester period is to identify early diagnostic or predictive molecules related to GDM disease. The primary clinical benefits of the research focused on the second and third trimesters of pregnancy include the ability to precisely assess the state of GDM, recognize distinct metabolic markers and metabolic pathways, forecast the course of pregnancy and the prognoses of the mother and fetus, or assess the effectiveness of medication or diet therapy (Li et al. 2019).

1.3 MACHINE LEARNING AND ITS APPLICATIONS IN HEALTHCARE

Humans comprehend the world by observing it and learning to anticipate future events. This involves repeatedly testing the model against observed data and making adjustments to enhance its accuracy. Similar to other multicellular organisms, humans are composed of numerous cells, which serve as the fundamental building blocks of life. In complex multicellular organisms, these cells are arranged into tissues, forming groups of similar cells that collaborate to perform specific functions. Organs consist of two or more tissues arranged to fulfill a specific function. Disease occurs when the regulatory processes malfunction or fail to function properly (Golemis et al. 2018). Disruption in the regulatory mechanisms can occur through various means, such as mutations in single nucleotides, insertions and deletions, alterations in copy numbers, modifications in epigenetic patterns, and modifications in the conformations of chromatin. Identifying the changes accountable for the disturbance in these biochemical processes is crucial for implementing an effective intervention. Recent progress in cutting-edge sequencing methods and diverse imaging techniques now enable comprehensive examination of genomic, epigenetic, and protein traits in both normal and diseased tissues. Collectively, these molecular profiles constitute the disease's genotype. The use of electronic health records is extensively integrated into clinical practices and research, offering a thorough depiction of the disease's phenotype (Overby 2013).

Researchers have long sought to comprehend complex biological systems, and recent advancements have facilitated significant progress in this endeavor. The continuously diminishing costs of high-throughput sequencing, the emergence of massively parallel technologies, and the innovation of new sensor technologies have empowered the generation of multidimensional data about biological systems. These dimensions encompass DNA sequence, epigenomic state, single-cell expression activity, proteomics, functional and phenotypic measurements, as well as ecological and lifestyle properties (The ENCODE Project Consortium 2012).

The term 'machine learning' broadly encompasses the process of constructing predictive models from data or identifying meaningful patterns within datasets. Essentially, ML seeks to replicate human pattern recognition abilities objectively through computational means. It proves particularly valuable when dealing with datasets that are either too extensive (containing numerous individual data points) or too intricate (featuring a high number of features) for human analysis. This is especially true when there is a need to automate data analysis for the establishment of a reproducible and time-efficient pipeline. Biological experiments often yield datasets with these characteristics, as they have substantially increased in size and complexity over the past few decades. It has become crucial not only to have practical methods for interpreting this wealth of data but also to possess a solid understanding of the employed techniques.

ML has emerged as a transformative force in healthcare, revolutionizing various aspects of medical research, diagnosis, and patient care. In diagnostics, ML algorithms have proven adept at analyzing complex medical data, including imaging and genomic information, to enhance accuracy in disease detection. Research highlights the successful positioning of ML models for the early identification of conditions such as cancer, cardiovascular diseases, and neurological disorders. These models not only exhibit superior diagnostic capabilities but also contribute to personalized treatment strategies by considering individual patient profiles. Furthermore, its predictive capabilities play a pivotal role in prognostics and risk assessment. Studies underscore the effectiveness of ML algorithms in forecasting patient outcomes, predicting disease progression, and optimizing treatment plans. This has profound implications for resource allocation, treatment optimization, and improving overall healthcare efficiency. In addition to diagnosis and prognosis, ML applications extend to drug discovery, personalized medicine, and healthcare management.

By making use of ML algorithms, it is possible to amalgamate extensive and diverse datasets, including clinical records, imaging data, and genomic information. This integration facilitates a deeper comprehension of the genetic underpinnings of diseases and aids in the identification of the most effective therapeutic strategies. Numerous research initiatives have been formulated to leverage this methodology, such as the All of Us research program. This program, to gather genomic sequence data from one million participants, aspires to play a pivotal role in advancing the precision medicine research platform.

This review provides an understanding of ML techniques, it aims to connect specific techniques with different types of biological data, drawing parallels with available reviews in specific biological disciplines. The abstracted research emphasizes the potential of ML to revolutionize healthcare delivery, and lead in an era of precision medicine and data-driven decision-making, ultimately leading to improved patient outcomes and enhanced healthcare systems.

1.3.1 Types of Machine Learning

There are three fundamental learning paradigms in ML. Supervised learning, unsupervised learning, reinforcement learning, and deep Learning (Table 1.2).

1.3.1.1 Supervised learning

In supervised learning, the process entails developing a predictor by utilizing training data composed of samples with identifiable class labels. In this approach, the goal is to predict the class of a new sample based on its known characteristics. For instance, the prognosis of a patient could be predicted using a model built from the gene expression patterns of tumor samples, incorporating data from tumor samples with known patient outcomes. In supervised learning, the procedure is termed "classification" when dealing with categorical output labels (such as good versus poor prognosis) and "regression" when handling continuous output (such as the number of months survived). Examples of supervised

TABLE 1.2 Different machine learning methods commonly used in healthcare.

Method	Description	Applications in Healthcare
Supervised learning	Uses labeled training data to make predictions	Diagnosing diseases (e.g., cancer, diabetes), predicting patient outcomes, drug discovery and development
Unsupervised learning	Learns from unlabeled data without predefined outputs	Identifying patterns in medical images, anomaly detection in healthcare data
Reinforcement learning	Learns through trial and error with rewards and penalties	Personalized treatment plans, clinical decision support systems
Deep learning	Uses deep neural networks to learn complex patterns, and medical image analysis (e.g., CT scans, MRI)	Natural language processing in healthcare, drug discovery and development

learning algorithms include linear and logistic regressions. The learned predictor, or more precisely, the predictive function, can take the form of a straight line or a highly intricate curve within a multidimensional space, contingent on the characteristics of the provided data. Numerous classification algorithms, totaling several hundred, have been developed (Fernandez et al. 2014). In supervised learning, a critical consideration is evaluating the classification model's performance when classifying a new sample. Consequently, a complex model may not only capture genuine patterns but also incorporate noise from the training data, leading to a phenomenon known as overfitting. Overfitting can result in a situation where the prediction error average during training is significantly lower than the error observed on unfamiliar samples. This application of Occam's razor helps mitigate overfitting. Cross-validation functions as a method for evaluating a model's ability to predict novel data not employed in its training phase. Its primary objective is the identification of potential issues such as overfitting or selection bias, providing insights into the model's capacity to generalize to an independent dataset (Varma and Simon 2006).

Supervised learning techniques have proven effective in developing prognostic and predictive biomarkers. A patient's clinical outcome is influenced by various factors, including inherent patient characteristics, the nature of the disease, and the impact of administered treatments. Prognostic factors, which are independent of treatment, play a role in patient outcomes, while biomarkers capable of identifying responses to specific therapeutic regimens are termed predictive biomarkers. Prognostic biomarkers aid in identifying patients requiring aggressive treatment, whereas predictive biomarkers assist in determining the optimal therapy for an individual.

Phase 1 studies represent the initial analyses aimed at generating hypotheses pinpointing potential markers for additional examination. Subsequently, Phase 2 studies extend the investigation and evaluate the association between a marker and prediction. Finally, Phase 3 studies, characterized by their extensive scale, serve as confirmatory investigations that explicitly articulate earlier hypotheses. Phase 3 studies must adhere to a well-defined protocol to ensure the attainment of the maximum level of evidence (Kambhampati and Mistry 2023).

Effective ML methods utilized in developing prognostic markers encompass regression models, classification, and regression trees.

1.3.1.2 Unsupervised Learning

It involves algorithms tasked with discovering patterns or structures within unlabeled data. Clustering and dimensionality reduction are common applications, allowing the algorithm to discern inherent relationships and groupings within the dataset. Cluster analysis is a technique that organizes samples into distinct groups based on the similarity between them. These groups consist of samples that are more alike than those in other groups. Various clustering algorithms, each with its advantages and disadvantages, are employed in biomedical research. Hierarchical clustering, self-organizing maps, and K-means are among the most frequently utilized algorithms for this purpose (D'haeseleer 2005). Unsupervised learning has found extensive application in the identification of homogeneous subgroups within biological samples. Given the heterogeneity observed in clinical outcomes and therapeutic responses across various diseases, it becomes crucial to discern homogeneous patient subgroups. This aids in pinpointing the biological factors influencing the disease and the efficacy of treatments. While this methodology has demonstrated significant success in treating individuals with breast cancer, there is a growing awareness that extending this approach to all types of cancer could be advantageous (Cancer Genome Atlas Network 2012).

1.3.1.3 Reinforcement Learning

This is centered around the concept of an agent interacting with an environment. The agent takes action, receives feedback in the form of rewards or penalties, and adjusts its strategy to maximize cumulative rewards over time. This paradigm is instrumental in training autonomous systems and decision-making processes. The use of reinforcement learning techniques has proven effective in optimizing antiretroviral therapy for human immunodeficiency virus (HIV) and in defining optimal strategies for managing sepsis. Reinforcement learning algorithms are employed to effectively learn optimal policies. Despite facing challenges, these algorithms are actively utilized to address significant issues within the healthcare sector (Gottesman et al. 2018).

1.3.1.4 Deep Learning

It is a subset of ML, which focuses on neural networks with multiple layers (deep neural networks). This architecture enables the automatic extraction of hierarchical features from data, contributing to remarkable achievements in image recognition, natural language processing, and other complex tasks (LeCun et al. 2015). Deep learning, also known as deep neural networks (DNN), has experienced substantial growth in recent years. DNNs belong to the category of artificial neural networks (ANNs) characterized by numerous concealed layers and distinct activation functions.

The fundamental unit of the ANN is a neuron, it accepts a collection of input values and generates a solitary output derived from them. This is typically accomplished by adding up the linear combination of all inputs, with each input multiplied by a corresponding weight, and subsequently applying the outcome to a non-linear function. A layer consists of a group of these neurons, and there are several layers, including hidden layers, positioned between the input and output layers. In the context of DNN, the term "deep" is used when there are multiple hidden layers, with the number exceeding one (Sahraeian et al. 2019). There have been notable advancements in the application of DNN in clinical contexts. For instance, deep learning was employed to directly predict the outcomes of anti-cancer drug combinations utilizing data on gene expression derived from experiments conducted on cell lines. Convolutional neural networks (CNNs) are extensively utilized in the analysis of clinical images, such as classifying skin lesions as malignant or benign, achieving accuracies comparable to dermatologists (Esteva et al. 2017). In a study, Khosravi et al. (2019) utilized a CNN structure to evaluate images of blastocysts in IVF instances, outperforming the predictive capabilities of individual embryologists in assessing the quality of embryos for successful pregnancies.

1.4 APPLICATION OF MACHINE LEARNING IN HEALTHCARE

ML has its roots in the 1950s, with Alan Turing introducing the concept of the first machine capable of learning and achieving artificial intelligence (Turing 1950). The latest developments in this field have displayed remarkable advancements, presenting opportunities to alleviate the workload of physicians and enhance the precision, forecasting, and quality of healthcare. ML is revolutionizing healthcare with advanced algorithms to extract meaningful insights from vast datasets. The integration of machine learning techniques into medical processes enhances diagnostics, treatment plans, and overall patient care. In diagnostics, ML excels at image and signal analysis, aiding in the detection of anomalies and diseases from medical imaging and diagnostic tests. The ability to analyze complex patterns in patient data allows for early disease detection and personalized treatment strategies. Present machine learning progress in the healthcare sector has predominantly functioned as a supportive tool, aiding physicians and analysts in fulfilling their responsibilities, recognizing trends in healthcare, and constructing models for predicting diseases. Within extensive medical institutions, strategies rooted in ML have been adopted to enhance efficiency in the management of electronic health records (Habehh and Gohel 2021) is conducted through medical imaging and monitoring. Additionally, ML is applied in robot-assisted surgeries (Kaouk et al. 2019). Recent applications of ML have facilitated the swift testing and hospital response in the fight against COVID-19. Through a deep learning system known as the Clinical Command Center developed by GE, hospitals have efficiently managed, shared, and tracked various resources such as patients, beds, rooms, ventilators, electronic health records (EHRs), and staff throughout the pandemic (Habehh and Gohel 2021). ML also contributes to predictive analytics, forecasting patient outcomes and identifying individuals at risk, thereby enabling proactive interventions. This proactive approach is particularly valuable in chronic disease management (Figure 1.2).

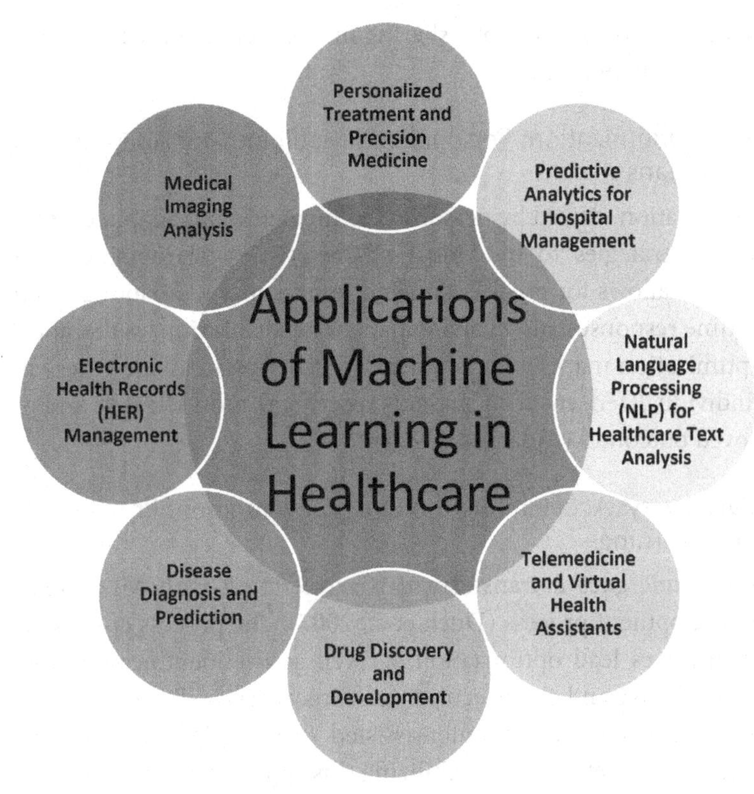

FIGURE 1.2 Various applications of machine learning in healthcare.

1.4.1 Diagnostics: Machine Learning Applications in Medical Imaging for Disease Detection

Integration of ML into medical imaging accelerates disease detection, transforming diagnostics. Algorithms analyze vast image datasets, enhancing accuracy and efficiency. In radiology, ML aids in early detection of abnormalities in X-rays, magnetic resonance imaging (MRIs), and computed tomography (CT) scans. Computer-aided detection systems improve diagnostic speed and precision, particularly in cancer screenings. These advancements not only expedite diagnosis but also contribute to personalized treatment plans, fostering a new era of more effective and targeted healthcare interventions. ML in medical imaging stands as a cornerstone in advancing diagnostics, promising improved patient outcomes and streamlined clinical workflows (Cury et al. 2021).

1.4.2 Prognostics: Predictive Modeling for Patient Outcomes and Disease Progression

Prognostics, fueled by predictive modeling, revolutionizes healthcare by forecasting patient outcomes and disease progression. Leveraging advanced algorithms on patient data provides clinicians with valuable insights for personalized treatment plans and early interventions. This approach not only enhances healthcare efficiency but also contributes to improved patient care by anticipating and mitigating potential complications.

Prognostics stands as a powerful tool, ushering in a new era of proactive and data-driven healthcare strategies (Freitas 2024).

1.4.3 Treatment Optimization: Personalized Medicine and Adaptive Treatment Plans

Treatment Optimization, driven by personalized medicine and adaptive treatment plans, tailors healthcare strategies to individual characteristics. Harnessing patient-specific data, this approach allows for precise, targeted interventions. Adapting treatment plans based on real-time responses and evolving patient needs maximizes therapeutic efficacy. Treatment Optimization marks a paradigm shift, ushering in an era where healthcare is increasingly individualized, ensuring patients receive the most effective and personalized care for improved outcomes (Ajdari et al. 2018).

1.4.4 Drug Discovery: Accelerating the Drug Development Process with Machine Learning

Drug Discovery undergoes a transformative acceleration with ML, streamlining the lengthy drug development process (Duch et al. 2007). The process of drug development and discovery involves lead optimization, as well as the identification and validation of targets, hit discovery, and the execution of clinical trials. To address this challenge, clinicians have employed the computer-assisted drug design (CADD) technique, as discussed by Hassan Baig et al. (2016). Utilizing this approach in drug discovery, artificial methods not only furnish molecular properties but also offer *in silico* lead compounds with desirable attributes. ML algorithms analyze vast biological datasets, predicting potential drug candidates and their efficacy. This data-driven approach expedites target identification, compound screening, and toxicity prediction. By significantly reducing time and resource demands, ML in drug discovery holds promise for faster, more cost-effective development of novel therapeutics, addressing critical healthcare needs, and advancing pharmaceutical innovation (Duch et al. 2007).

1.5 CONCLUSION

In summary, the convergence of ML and healthcare has yielded profound advancements, revolutionizing diagnostics, treatment strategies, and patient outcomes. The robust predictive capabilities of ML algorithms have demonstrated substantial efficacy in clinical decision-making and risk assessment. Challenges persist, including the need for standardized data and robust validation methodologies. The symbiotic relationship between computational methodologies and medical sciences necessitates ongoing interdisciplinary collaboration. As ML applications continue to evolve, refining interpretability and ensuring ethical considerations will be paramount. The trajectory of ML in healthcare promises a paradigm shift, underscoring the imperative for rigorous scientific inquiry and continual refinement to fully unlock its transformative potential. The integration of systems biology and ML in reproductive health shows potential for understanding intricate molecular interactions that contribute to fertility and pregnancy outcomes. This synergy can enable the development of prediction models, customized therapies, and innovative interventions, propelling reproductive medicine toward more accurate and efficient solutions.

REFERENCES

Ajdari A, Ghate A, Kim M (2018) Adaptive treatment-length optimization in spatiobiologically integrated radiotherapy. *Phys Med Biol* 63:075009. https://doi.org/10.1088/1361-6560/aab4b6

Ashburner M, Ball CA, Blake JA, et al (2000) Gene ontology: tool for the unification of biology. *Nat Genet* 25:25–29. https://doi.org/10.1038/75556

Birney E (2021) The International Human Genome Project. *Hum Mol Genet* 30:R161–R163. https://doi.org/10.1093/hmg/ddab198

Cantonwine DE, Hauser R, Meeker JD (2013) Bisphenol A and human reproductive health. *Exp Rev Obstet Gynecol* 8:329–335. https://doi.org/10.1586/17474108.2013.811939

Chang K-J, Seow K-M, Chen K-H (2023) Preeclampsia: recent advances in predicting, preventing, and managing the maternal and fetal life-threatening condition. *IJERPH* 20:2994. https://doi.org/10.3390/ijerph20042994

Costa PR, Acencio ML, Lemke N (2010) A machine learning approach for genome-wide prediction of morbid and druggable human genes based on systems-level data. *BMC Genom* 11:S9. https://doi.org/10.1186/1471-2164-11-S5-S9

Cury RC, Megyeri I, Lindsey T, et al (2021) Natural language processing and machine learning for detection of respiratory illness by chest CT imaging and tracking of COVID-19 pandemic in the United States. *Radiol Cardiothorac Imag* 3:e200596. https://doi.org/10.1148/ryct.2021200596

D'haeseleer P (2005) How does gene expression clustering work? *Nat Biotechnol* 23:1499–1501. https://doi.org/10.1038/nbt1205-1499

Ding X, Schimenti JC (2021) Strategies to identify genetic variants causing infertility. *Trends Mol Med* 27:792–806. https://doi.org/10.1016/j.molmed.2020.12.008

Duch W, Swaminathan K, Meller J (2007) Artificial intelligence approaches for rational drug design and discovery. *CPD* 13:1497–1508. https://doi.org/10.2174/138161207780765954

Dulbecco R (1986) A turning point in cancer research: sequencing the human genome. *Science* 231:1055–1056. https://doi.org/10.1126/science.3945817

Esteva A, Kuprel B, Novoa RA, et al (2017) Dermatologist-level classification of skin cancer with deep neural networks. *Nature* 542:115–118. https://doi.org/10.1038/nature21056

Fernandez-Delgado M, Cernadas E, Barro S, Amorim D (2014) *Do we Need Hundreds of Classifiers to Solve Real World Classification Problems?* https://jmlr.org/papers/v15/delgado14a.html

Freitas, AT (2024) Data-driven approaches in healthcare: Challenges and emerging trends, in: Sousa Antunes H, Freitas PM, Oliveira AL, Martins Pereira C, Vaz De Sequeira E, Barreto Xavier L (eds.). *Multidisciplinary Perspectives on Artificial Intelligence and the Law.* Springer International, Cham, pp. 65–80. https://doi.org/10.1007/978-3-031-41264-6_4.

Golemis EA, Scheet P, Beck TN, et al (2018) Molecular mechanisms of the preventable causes of cancer in the United States. *Genes Dev* 32:868–902. https://doi.org/10.1101/gad.314849.118

Gottesman O, Johansson F, Meier J, et al (2018) Evaluating reinforcement learning algorithms in observational health settings. arXiv:1805.12298 cs.LG. https://doi.org/10.48550/ARXIV.1805.12298

Habehh H, Gohel S (2021) Machine learning in healthcare. *CG* 22:291–300. https://doi.org/10.2174/1389202922666210705124359

Hansen J, Meretzky D, Woldesenbet S, et al (2017) A flexible ontology for inference of emergent whole cell function from relationships between subcellular processes. *Sci Rep* 7:17689. https://doi.org/10.1038/s41598-017-16627-4

Hassan Baig M, Ahmad K, Roy S, et al (2016) Computer aided drug design: success and limitations. *CPD* 22:572–581. https://doi.org/10.2174/1381612822666151125000550

Jeon J, Nim S, Teyra J, et al (2014) A systematic approach to identify novel cancer drug targets using machine learning, inhibitor design and high-throughput screening. *Genome Med* 6:57. https://doi.org/10.1186/s13073-014-0057-7

Kambhampati R, Mistry K (2023) Elements of clinical trial protocol design. In: Jagadeesh G, Balakumar P, Senatore F (eds). *The Quintessence of Basic and Clinical Research and Scientific Publishing* [Internet]. Singapore: Springer Nature Singapore; p. 361–71. Available from: https://link.springer.com/10.1007/978-981-99-1284-1_22 (accessed on Sep 25, 2024)

Kandoi G, Acencio ML, Lemke N (2015) Prediction of druggable proteins using machine learning and systems biology: a mini-review. *Front Physiol* 6:366. https://doi.org/10.3389/fphys.2015.00366

Kaouk JH, Garisto J, Eltemamy M, Bertolo R (2019) Robot-assisted surgery for benign distal ureteral strictures: step-by-step technique using the SP® surgical system. *BJU Intl* 123:733–739. https://doi.org/10.1111/bju.14635

Khosravi P, Kazemi E, Zhan Q, et al (2019) Deep learning enables robust assessment and selection of human blastocysts after in vitro fertilization. *npj Digit Med* 2:21. https://doi.org/10.1038/s41746-019-0096-y

Kitano H (2002) Systems biology: a brief overview. *Science* 295:1662–1664. https://doi.org/10.1126/science.1069492

LeCun Y, Bengio Y, Hinton G (2015) Deep learning. *Nature* 521:436–444. https://doi.org/10.1038/nature14539

Li G, Gao W, Xu Y, et al (2019) Serum metabonomics study of pregnant women with gestational diabetes mellitus based on LC-MS. *Saudi J Biol Sci* 26:2057–2063. https://doi.org/10.1016/j.sjbs.2019.09.016

Ma'ayan A (2017) Complex systems biology. *J R Soc Interface* 14:20170391. https://doi.org/10.1098/rsif.2017.0391

Odenkirk MT, Stratton KG, Gritsenko MA, et al (2020) Unveiling molecular signatures of preeclampsia and gestational diabetes mellitus with multi-omics and innovative cheminformatics visualization tools. *Mol Omics* 16:521–532. https://doi.org/10.1039/D0MO00074D

Overby CL, Kohane I, Kannry JL, Williams MS, Starren J, Bottinger E, et al. (2013) Opportunities for genomic clinical decision support interventions. *Genet Med* 15:817–23. https://doi.org/10.1038/gim.2013.128.

Sahraeian SME, Liu R, Lau B, et al (2019) Deep convolutional neural networks for accurate somatic mutation detection. *Nat Commun* 10:1041. https://doi.org/10.1038/s41467-019-09027-x

Spanou L, Dimou A, Kostara CE, et al (2022) A study of the metabolic pathways affected by gestational diabetes mellitus: comparison with type 2 diabetes. *Diagnostics* 12:2881. https://doi.org/10.3390/diagnostics12112881

Tavassoly I, Goldfarb J, Iyengar R (2018) Systems biology primer: the basic methods and approaches. *Essays Biochem* 62:487–500. https://doi.org/10.1042/EBC20180003

The Cancer Genome Atlas Network (2012) Comprehensive molecular characterization of human colon and rectal cancer. *Nature* 487:330–337. https://doi.org/10.1038/nature11252

The ENCODE Project Consortium (2012) An integrated encyclopedia of DNA elements in the human genome. *Nature* 489:57–74. https://doi.org/10.1038/nature11247

Turing AM (1950) Computing machinery and intelligence. *Mind* LIX :433–460. https://doi.org/10.1093/mind/LIX.236.433

Varma S, Simon R (2006) Bias in error estimation when using cross-validation for model selection. *BMC Bioinform* 7:91. https://doi.org/10.1186/1471-2105-7-91

Wang Q-Y, You L-H, Xiang L-L, et al (2021) Current progress in metabolomics of gestational diabetes mellitus. *WJD* 12:1164–1186. https://doi.org/10.4239/wjd.v12.i8.1164

Yao L, Rzhetsky A (2008) Quantitative systems-level determinants of human genes targeted by successful drugs. *Genome Res* 18:206–213. https://doi.org/10.1101/gr.6888208

Zhao C, Ge J, Li X, et al (2020) Integrated metabolome analysis reveals novel connections between maternal fecal metabolome and the neonatal blood metabolome in women with gestational diabetes mellitus. *Sci Rep* 10:3660. https://doi.org/10.1038/s41598-020-60540-2

Zhu Y, Barupal DK, Ngo AL, et al (2022) Predictive metabolomic markers in early to mid-pregnancy for gestational diabetes mellitus: a prospective test and validation study. *Diabetes* 71:1807–1817. https://doi.org/10.2337/db21-1093

CHAPTER 2

Data Sources and Data Integration in Reproductive Health

Souradip Basu, Saptarshi Bhattacharyya, Saptaki De, Mahashweta Mitra Ghosh, and Sayak Ganguli

2.1 INTRODUCTION

The condition of the male and female reproductive systems across the lifespan is encapsulated by the term reproductive health. These systems comprise of organs and glands that produce hormones, with the pituitary gland in the brain being one such example. The female ovaries and male testicles serve as reproductive organs (gonads) crucial for maintaining the health of their respective systems. Additionally, these organs function as glands by producing and releasing hormones. Issues related to the female and male reproductive system encompass early or delayed onset of puberty;

- Endometriosis is characterized by the abnormal growth of endometrial tissue outside the uterus
- Inadequate supply of breast milk
- Challenges with fertility, including difficulties in conceiving or low fertility
- Menstrual irregularities, such as excessive or unpredictable bleeding
- Polycystic ovarian syndrome results in elevated production of male hormones by the ovaries.
- Complications during pregnancy
- Uterine fibroids are benign growths occurring in the uterus or womb of a woman.
- Erectile dysfunction or impotence
- Decreased sperm count

DOI: 10.1201/9781003487548-2

Environmental factors are believed by scientists to contribute to various reproductive issues. As per research findings, the impact of environmental influences on reproductive health includes:

- Lower fertility in both men and women has been linked to exposure to lead (Clay et al. 2018).
- Mercury exposure has been associated with difficulties in the nervous and the reproductive system, affecting aspects such as memory, attention, fertility and fine motor abilities (Santos-Lima et al. 2020).
- The use of diethylstilbestrol (DES), a medication administered to pregnant women, can elevate the risk of cancer, infertility, and pregnancy complications in their daughters (Titus et al. 2019).
- Exposure to substances that disrupt endocrine function, affecting hormones, can lead to problems related to puberty, fertility, and pregnancy (Li et al. 2015).

One of the prime organizations involved in the studies on reproductive health has been the Centers for Disease Control and Prevention (CDC), which has taken several key initiatives that have been instrumental in the holistic improvement of reproductive health research around the world. CDC's Division of Reproductive Health and its numerous initiatives have been depicted in Figure 2.1.

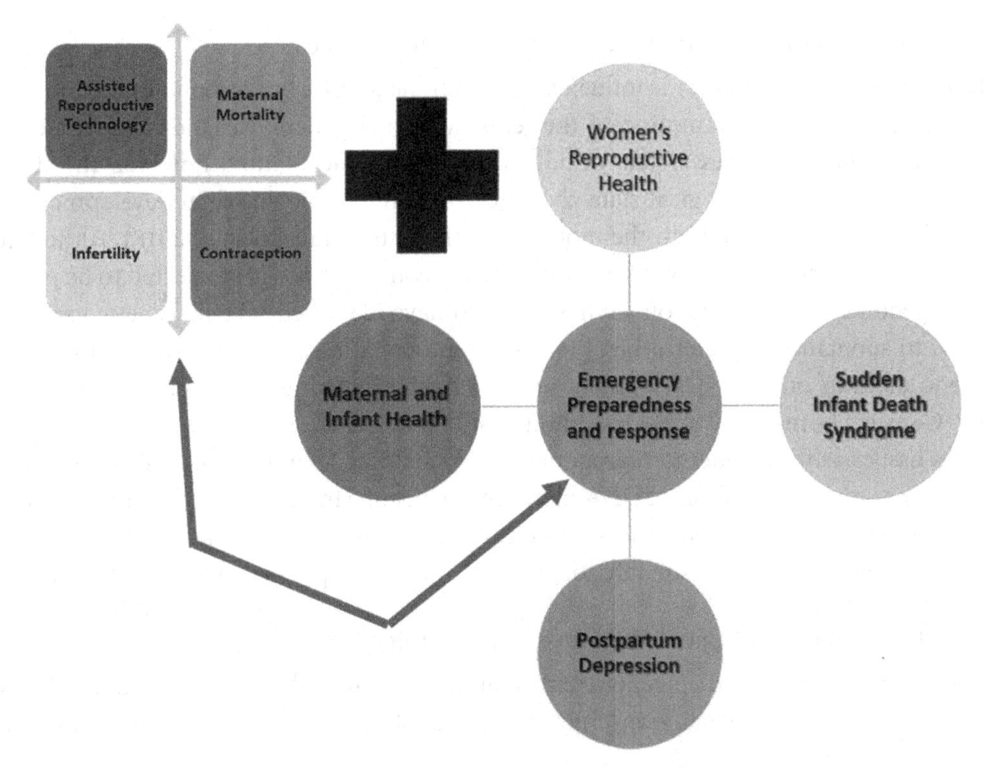

FIGURE 2.1 CDC's Division of Reproductive Health and its diverse initiatives.

2.1.1 Commonly Used Data Sources

Table 2.1 describes commonly used sources, which are enlisted along with acronyms.

2.2 IDENTIFYING ISSUES IN NEONATAL AND MATERNAL HEALTH

2.2.1 Role of Clinical and Administrative Data

Various sources of population-level data were employed to identify factors impacting the health of both mothers and children. One prominent source is electronic health records (EHR), which encompass demographics, anthropometrics, diagnoses, vital signs, medications, and diverse clinical data, offering a comprehensive overview of a patient's health history. Operational insurance claim data, known for their large sample sizes across wide geographic regions albeit with less detailed information, have demonstrated efficacy in constructing and analyzing cohorts related to pregnancy (Marić et al. 2020). Research leveraging EHR-based models to forecast conditions such as gestational diabetes mellitus (GDM), preeclampsia, and preterm birth (PTB) underscores the importance of these data sources in the field of reproductive health. Furthermore, markers like proteinuria and bacteriuria, identified through urinalysis and urine culture, serve as dependable predictors for severe conditions such as preeclampsia (Tanaçan et al. 2019). Recent advancements in artificial intelligence (AI) have enabled clinicians to incorporate these vital, cost-effective tests into more advanced predictive and diagnostic models than were previously achievable (Burton et al. 2019).

2.2.2 Insights from Epigenetics and Genetic Data Sources

The occurrence of pregnancy is influenced by both the genetic background of the mother and the epigenetic programming of the fetus. Maternal genetic variations impact crucial developmental processes like decidualization and placentation; however, the delicately balanced immunological state characterizing embryonic and fetal development is shaped by the genetics of both the mother and the fetus (Huusko et al. 2018). Although the genetic factors influencing different pregnancy complications are expected to be polygenic, genome-wide association studies (GWAS) have identified specific genetic variants linked to spontaneous preterm birth (sPTB), demonstrating the influence of a mother's genetic makeup on fetal development and pregnancy outcomes even prior to conception. Conversely, maternal epigenetic modifications reflect the status of various cell types across tissues as they undergo reprogramming to support fetal growth. Evaluating these alterations—through analysis of DNA methylation, histone markers, and chromatin accessibility in the relevant cell subsets—offers insights into whether specific cell types at the maternal–fetal interface are suitably adapted for a successful pregnancy (Kwak et al. 2019).

2.2.3 Elucidating Biological Pathways Using Omics Data

In addition to genetic and epigenetic information, employing high-throughput methods to investigate the maternal transcriptome, metabolome, proteome, and microbiome can offer a more comprehensive understanding of the biological status of the mother. Given the extensive alterations in tissue-specific gene expression in maternal tissues during pregnancy, coupled with concurrent metabolic, hormonal, and immunological adaptations, diverse modalities can unveil various facets of maternal and fetal health at specific time points (Handelman et al. 2019). For instance, cell-free RNA (cfRNA) extracted from blood or amniotic fluid can be employed to monitor the dynamics of fetal development and placentation, offering implications for the early detection of pre-eclampsia by identifying markers of impaired placentation in maternal plasma (Park et al. 2021). Microbiomics, conversely, can explore the vaginal, oral, and gut microbiomes, all of which are vital in preserving the immunological equilibrium required for fetal tolerance (Mei et al. 2019). Disruptions in the vaginal microbiome, as evidenced in bacterial vaginosis, can potentially result in PTB due to inflammatory mechanisms or intra-amniotic infections prompted by vaginal microorganisms. Likewise, pathogenic oral microorganisms can exacerbate PTB by inducing systemic inflammation stemming from oral ailments or hematogenous spread to the amniotic fluid (Cobb et al. 2017). Lastly, the gut microbiome, which influences the metabolism of small molecules and hormones while also regulating metabolic and immunological functions, has been associated with GDM and PTB when its equilibrium is disturbed (Yang et al. 2020). In summary, a variety of omics technologies possess substantial promise for elucidating the molecular constituents of pregnancy-related disorders.

2.3 USE OF MODELLING USING MACHINE LEARNING

2.3.1 Machine Learning

Machine learning encompasses diverse statistical techniques employed to draw inferences from data based on a set of attributes (e.g., clinical measurements of patients) (Wainberg et al. 2018). Supervised learning entails algorithms that learn to forecast a particular attribute or outcome (e.g., PTB) associated with a given set of features. It is labeled "supervised" because the outcome is known for a sufficiently large dataset (number of patients), facilitating the identification of a relationship between the features and the outcome, thereby enabling the development of a predictive model. In contrast, unsupervised learning lacks prior knowledge of the outcome and seeks to uncover concealed patterns or structures within the data (Wang 2001). Semi-supervised learning merges elements of both supervised and unsupervised learning, where the outcome is known for a limited number of samples. In the realm of pregnancy, the primary emphasis is on supervised learning, which involves utilizing information such as medical history or measurements from an omics assay (input) to forecast outcomes like gestational age (GA) or PTB (output). Usually, the input is structured in a tabular format, where each row denotes a sample, the columns represent categorical or numeric data, and the output is a single outcome.

TABLE 2.1 Some existing data sources for studies in reproductive health.

S. No.	Title	Purpose	Sponsorship	Design
1	The National Survey of Family Growth (NSFG)	"The National Survey of Family Growth (NSFG) is a vital resource for comprehensively understanding multiple facets of fertility within the United States. It provides invaluable insights into fertility trends, infertility rates, reproductive health, contraception practices, and fertility intentions. Furthermore, the survey covers a wide array of topics related to child development, including unwanted childbearing, adoption, adolescent pregnancy, single motherhood, prenatal and postnatal care, as well as infant health. The data collected by the NSFG can be analyzed alongside various social, economic, and familial factors, enriching the depth of research findings. Given its extensive history since 1955, the survey facilitates the examination of time-series statistics, allowing researchers to track changes in family planning, contraceptive usage, and reproductive health patterns over time. Researchers, healthcare professionals, demographers, social scientists, and policymakers at both the national and local levels rely on the NSFG data for conducting diverse studies and making informed policy decisions aimed at addressing reproductive health needs and challenges within the population.	The National Survey of Family Growth is administered by the Family Growth Survey Branch within the Division of Vital Statistics, which operates under the auspices of the National Center for Health Statistics. Financial backing for the survey is sourced from multiple entities, including the Office of Family Planning Services within the former Bureau of Community Health Services, the Center for Population Research at the National Institute of Child Health and Human Development (NICHD), the Office of Adolescent Programs, and the National Center for Health Statistics (NCHS)	The National Survey of Family Growth (NSFG) administers interviews to women aged 15–44, encompassing all marital statuses, in a manner that ensures national representation. The 1982 survey involved a sample of approximately 8,000 women selected through area probability sampling, with a deliberate oversampling of 1,900 teenagers. Parental consent is obtained for all minor participants. Two distinct questionnaires are designed for women under 25 and those aged 25 and older, respectively. Significantly, the 1982 interview, known as Cycle III of the NSFG, marked the first time all women within the childbearing age range were included, regardless of their marital status. To facilitate specific analyses, there was a deliberate oversampling of Black participants

Periodicity	Content	Availability

(*continued*)

TABLE 2.1 (Continued)

S. No.	Title	Purpose	Sponsorship	Design
2	National Longitudinal Surveys of the Labor Market Experience of: Young Women, Young Men, Mature Women, and Mature Men	This longitudinal survey series was initiated to examine the evolving labor market experiences of distinct cohorts facing employment challenges that are of particular interest to policymakers. These challenges include the transition from education to employment, initial career decisions, adaptation to workforce demands, balancing work and family responsibilities, and attaining stable employment. These issues are especially relevant for two specific cohorts: young men aged 14–24 in 1966 and young women aged 14–24 in 1968. Additionally, middle-aged men aged 45–59 in 1966 face concerns related to declining health, unemployment, skill obsolescence, and age-related discrimination. Furthermore, for women aged 30–44 in 1967, the focus initially revolves around re-entering the labor force as their children grow older, with later emphasis shifting toward retirement issues. The longitudinal tracking of these cohorts enables analysts to not only depict the circumstances of diverse population groups but also understand the factors preceding and following situations involving education, employment, marriage, family dynamics, and economic status.	The Office of Manpower Policy Evaluation and Research at the Department of Labor spearheaded the launch of these four longitudinal surveys. The questionnaires were developed by the Center for Human Resource Research at Ohio State University, which also supplies computer tapes and comprehensive documentation. Fieldwork for the surveys is conducted by the U.S. Bureau of the Census.	Each of the four age–sex cohorts is represented by a multi-stage probability sample. To ensure statistically reliable statistics for Black participants, households in predominantly Black enumeration districts were sampled at a rate between three and four times that of other households. From over 35,000 inhabited housing units, a sample of 5,050 men aged 45–59 was interviewed. A sample of 5,225 males aged 14–24, excluding those on active military service, was also interviewed. Additionally, 5,083 women aged 30–44 and 5,159 young women aged 14–24 were interviewed. Several families in the sample have contributed more than one respondent because there are a total of 13,582 households among the four NLS samples. While most interviews were initially conducted in-person, most interviews during the 1970s were conducted over the phone. The data are nationally representative when weighted to account for sample attrition and oversampling. Approximately 56% of the males initially aged 45–59, 65% of the younger men, and approximately 70% of the two women's cohorts who were first interviewed were still being interviewed as of the 15-year interview points.

Periodicity	Content	Availability
Every year from 1968 to 1973, as well as in 1975, 1977, 1978, 1980, 1982, 1983, and 1985, interviews with young women were conducted. There will be more interviews in 1987 and 1988. Every year from 1967 to 1969, as well as in 1971–1972, 1974–1976, 1977, 1979, 1981–1982, and 1984, women were questioned. We have scheduled interviews for 1986 and 1987. Interviews with young men took place every year from 1966 to 1971, as well as in 1975, 1976, 1978, 1980, and 1983. There will be no further interviews. Between 1966 and 1969, as well as in 1971, 1973, 1975, 1976, 1978, 1980, 1981, and 1983, men were interviewed on a yearly basis. There will be no further interviews.	According to the surveys' focus on labor force concerns, a large number of questions center on employment experience, unemployment, income, and training. Nonetheless, a great deal of data regarding the family's history as well as their social and economic standing was gathered. After the mid-1970s, none of the respondents were still underage; yet, by the 1980s, most of the young men and women had become parents, and there is some scant data on their offspring. A great deal of data about the familial circumstances of the young men and women who responded as children was gathered; it is displayed below	From the Center for Human Resource Research, 5701 North High Street, Worthington, Ohio 43085, you can obtain data cassettes, full documentation, and a list of publications. Contact the appropriate cohort coordinators or Pat Rhoton: Mature men—Gilbert Nestel; Mature women—Lois Shaw; Young men—Stephen Hills; Young women—Frank Mott, or Principal; Investigator for the NLS—Ken Wolpin (514); 888-8238 or (614) 422-7337.

(*continued*)

TABLE 2.1 (Continued)

S. No.	Title	Purpose	Sponsorship	Design
3	Current Population Survey-Fertility Supplements	The purpose of the fertility supplements is to give national estimations of a woman's fertility and her anticipated future birth rate. Furthermore, certain additions (1977, 1982) have offered details regarding the childcare arrangements employed by working moms for their youngest kid who is younger than5 years old	The U.S. Bureau of the Census is solely responsible for the supplement's sections on fertility and birth expectations. The Department of Health and Human Services funded the childcare sections of the 1977 supplement, and the Bureau and the National Institute of Child Health and Human Development jointly sponsored an extended fertility supplement in 1980	The report of the core survey included a description of the fundamental layout of the Current Population Survey. All individuals living in sampling households who satisfy specific eligibility requirements have been asked the additional questions. These requirements have most recently been limited to women who are single and between the ages of 15 and 59 or 18 and 59. Women aged 18–44 is questioned about their plans for having children. These age requirements, however, have ranged from 14 to 75 years old. Data on men's and women's marriage histories were collected for the enlarged fertility supplement for ages 15–75
4	The National Surveys of Young Women and Men (Kantner-Zelnik data)	The main source of information on the sexual experiences of American women in their 15–19-year-old age range throughout the 1970s and men in their 17–21-year-old age range in 1979 has been the Kantner–Zelnik study. The three surveys conducted in 1971, 1976, and 1979 also gathered data on the use of contraceptives, pregnancies, intention to become pregnant, and experiences with sex education.	These surveys have been mostly conducted by Melvin Zelnik and John I. Kantner. The General Services Foundation, the Ford Foundation, NICHD, and the Center for Population Research have all contributed funding.	For the three interviews, the designs have somewhat changed. In the 1971 survey, women aged 15–19 who lived in families across the continental United States were questioned (N = 4611). Additionally, young women in college dorms were interviewed (N = 219), and a second sample of young women in college dormitories was also used. A sample of 2,500 women, aged 15–19, who were living in households in the continental United States and were born between March 1956 and February 1961 were included in

Periodicity	Content	Availability
Since 1971, the addition has been carried out in June each year. For 1984, there will be an addition	Data on fertility and expected birth weights are gathered for every supplement. Furthermore, statistics on child spacing and marriages are included in the 1971, 1975, and 1980 supplements; data on childcare is included in the 1977 and 1982 supplements. For the first time, information about men's prior marriages' children under the age of 18 and if any of them reside elsewhere were added in the 1980 supplement, which gathered data on both men's and women's marriage histories.	Only the total number of births, the date of the youngest child's birth (and occasionally the eldest child's birth), and the number of projected additional children are provided in the typically brief supplement. Only working moms with children under five and the youngest of these children's daycare needs are covered by the childcare parts of the 1977 and 1982 laws. The type of payment (cash or non-cash) is recorded, but not the total amount. No information on unwed births to younger teenagers is available from this source because unmarried women under the age of 18 are excluded from any of the supplements. Retrospective histories of this kind are prone to significant mistake, particularly when it comes to events that occurred years ago, as demonstrated by analyses of data from the marriage histories. Without a doubt, this effect is amplified by the survey's method of gathering data from proxy respondents. The facts pertaining to men are the most severely impacted, as the majority of respondents are women. The number of men's children from prior marriages living abroad is far too low, according to comparisons with other data sources.
Three distinct years have seen the conduct of interviews: 1971, 1976, and 1979. In every cohort, there have been distinct respondents	Comprehensive information is gathered regarding sexual behavior, usage of contraceptives, intention to become pregnant, and prior sex education. Additionally, some background data was gathered	Abortions, pregnancies and births are underreported

(*continued*)

TABLE 2.1 (Continued)

S. No.	Title	Purpose	Sponsorship	Design
				the 1976 study. In the 1979 study, young men and women who were residing in families in the continental United States' Standard Metropolitan Statistical Areas (SMSAs) were included. The eligible male respondents (total N = 917) and female respondents (total N = 1,717) were born between March 1957 and February 1962 (ages 15–19).
5	Alan Guttmacher Institute (AGI)	A major source of information about abortion services in the United States is the Alan Guttmacher Institute. Since 1973, the AGI has conducted annual surveys of all recognized abortion providers in every state	Numerous private foundations provide financing to the Alan Guttmacher Institute	All states identified abortion providers are polled

Modified from National Research Council (US) Panel (1987).

2.3.2 Computational Modeling of Pregnancy Outcomes Using High-Throughput Biological Data

The vast amount of data generated by high-throughput multi-omics poses significant challenges for even the most advanced machine learning and statistical tools. These challenges stem from two key characteristics: First, the data often has a large number of features (high dimensionality) but a limited number of samples (e.g., patients). Second, the different types of omics data are inherently heterogeneous, making their integration, even without considering high dimensionality, a complex task. To analyze data with such high dimensionality and a limited sample size, researchers generally employ two main approaches: feature reduction and sparse modeling (Saeys et al. 2007). Feature reduction typically involves a preprocessing step that can either select significant features, filter out redundancies or compress the information into a smaller set of representative features through dimensionality reduction techniques like principal component analysis (PCA) (Cunningham and Ghahramani 2015). This helps make the data more manageable for further analysis. Sparse modeling, on the other hand, focuses on building models that rely on only a small subset of the features, leading to more interpretable results and improved generalizability to unseen data. Analyzing the extensive omics data generated during

Periodicity	Content	Availability
The survey's coverage spans the years 1973–1982	The Centers for Disease Control (CDC) provide data on several demographic factors including age, race, marital status, education, number of children, gestation at abortion, number of prior abortions, and abortion procedure. This information is combined with data from the Alan Guttmacher Institute (AGI) regarding the total number of abortions to generate national estimates	

pregnancy presents unique challenges due to its high dimensionality (many features) and limited sample size. To overcome these hurdles, researchers often employ two main approaches:

- **Feature Reduction:** This technique aims to simplify the data and minimize noise, often for visualization purposes. It can be accomplished through either feature selection, which identifies and retains only the most relevant features, or dimensionality reduction, which condenses the information into a smaller set of representative features. PCA is a common example of dimensionality reduction. Feature reduction is particularly beneficial for constructing interpretable models and enhancing data visualization.
- **Sparse Modeling:** This approach directly models an outcome while extracting the most informative features associated with that outcome. These features are then integrated into a prediction model that is both interpretable and computationally efficient. Sparse modeling also enables the integration of prior knowledge (a priori information) to further refine and streamline the models (Culos et al. 2020).

2.4 USE OF TIME SERIES DATA

Clinical time series data play a pivotal role in monitoring patients in real-time settings like the intensive care unit (ICU) and retrospective scenarios for prolonged monitoring and diagnoses. In pregnancy, fetal heart rate analysis is a relevant example used to detect pathological fetal conditions such as perinatal hypoxia, fetal growth restriction (FGR), and fetal arrhythmias and heart anomalies (Marzbanrad et al. 2018). Typically, practitioners extract various quantitative parameters related to fetal conditions, such as short-term variability (STV) and long-term irregularity (LTI), from fetal heart rate measurements. These parameters are either directly interpreted or transformed into a tabular format suitable for machine learning applications (Signorini et al. 2020). Indeed, advanced machine learning techniques, including deep learning and recurrent neural networks, hold promise in directly modeling and predicting pathological fetal conditions (Fawaz et al. 2019). These techniques leverage the power of AI to analyze complex data patterns and provide more accurate predictions compared to traditional methods.

As an example, deep learning can be employed directly on cardiotocography (CTG) recordings for fetal acidemia prediction, eliminating the need for manual feature extraction like STV and LTI markers. In this approach, deep learning architectures are tailored to analyze raw data, potentially revealing previously unknown markers to clinicians. The application of sensors and actigraphy to observe sleep quality, activity patterns, and movement in individuals presents a novel method for measuring lifestyle-related behaviors that could affect health (Souza et al. 2019). Although this method has received limited attention, research suggests that changes in sleep patterns during pregnancy may result in varying gene expression in mothers (Carroll et al. 2020). Machine learning methods designed for time series data analysis can be utilized to examine smartwatch and actigraphy data, offering valuable insights into different pregnancy-related parameters. Moreover, given the widespread use of smartphones and wearable gadgets, a broad array of techniques for activity recognition and semantic behavior analysis are now accessible, allowing for a more comprehensive exploration of potential behavioral risk factors (Nweke et al. 2018).

2.5 ADVANCEMENTS IN IMAGING TECHNOLOGIES

Image and video analyses, particularly employing machine learning and deep learning techniques, have exhibited significant potential across various domains (Liu et al. 2019). Doppler ultrasound, capable of measuring fetal and umbilical blood flow, has been investigated for its potential to predict adverse pregnancy outcomes during the initial and subsequent trimesters. Nonetheless, images and videos are frequently transformed into manually examined (handcrafted) features or inputted into machine learning models utilizing diverse feature extraction approaches. Both basic science and clinical research are increasingly adopting multiplexed imaging methods with ultrasound-related modalities, a crucial element in clinically evaluating fetal health, serving as a prominent example. Doppler ultrasound, with the ability to quantify fetal and umbilical blood flow, has been explored for predicting adverse pregnancy outcomes in the first and second trimesters. However, images and videos often undergo conversion into manually inspected features or are input into machine learning models using various feature extraction methods (Baharlou et al. 2019).

2.6 A HOLISTIC VIEW ON PREGNANCY

The burgeoning field of machine learning offers powerful tools for unraveling the complexities of pregnancy and its associated pathologies. Recent advancements in multimodal learning and multitask learning hold immense promise for capturing a more complete picture, as outlined in the accompanying Clinician's Corner.

2.6.1 Multimodal Learning: Blending Perspectives for Deeper Insights

This approach integrates data from multiple sources and modalities, such as genes, proteins, metabolites, clinical measurements, and even imaging data. By harnessing the strengths of each modality, multimodal models can build a richer understanding of biological processes in pregnancy, going beyond the limitations of any single data type. Imagine combining regular clinical data with time series measurements, ultrasound images, and even patient narratives to create a comprehensive model for predicting GA or identifying risks for complications. This holistic view holds the potential for more accurate predictions and personalized care (Li et al. 2019).

2.6.2 Multitask Learning: Uncovering Shared Secrets of Pregnancy Outcomes

Pregnancy outcomes are often intertwined, with different clinical manifestations potentially sharing underlying biological pathways. Multitask learning capitalizes on this interconnectedness by training models on multiple, related outcomes simultaneously. Instead of treating each outcome in isolation, this approach seeks solutions that capture shared patterns and biological structures across them, leading to more robust and generalizable models (Harutyunyan et al. 2019). Think of simultaneously predicting GA, preeclampsia risk, and FGR—the model may uncover shared genetic variants or metabolic pathways relevant to all three outcomes, providing deeper insights into their common threads.

2.6.3 The Synergy of Multimodal and Multitask Approaches

Combining the strengths of both paradigms creates a truly transformative perspective on pregnancy. Multimodal multitask learning allows for condensed representations of both the diverse data inputs and the interconnected outcomes. This compact and insightful representation promises a novel understanding of the complex interplay between genetic, molecular, and environmental factors that shape pregnancy and its various potential pathways. In essence, it opens the door to a holistic view of pregnancy, paving the way for personalized risk prediction, early intervention strategies, and ultimately, improved maternal and fetal health outcomes. By embracing these innovative machine learning tools, researchers can move beyond traditional, siloed approaches toward a

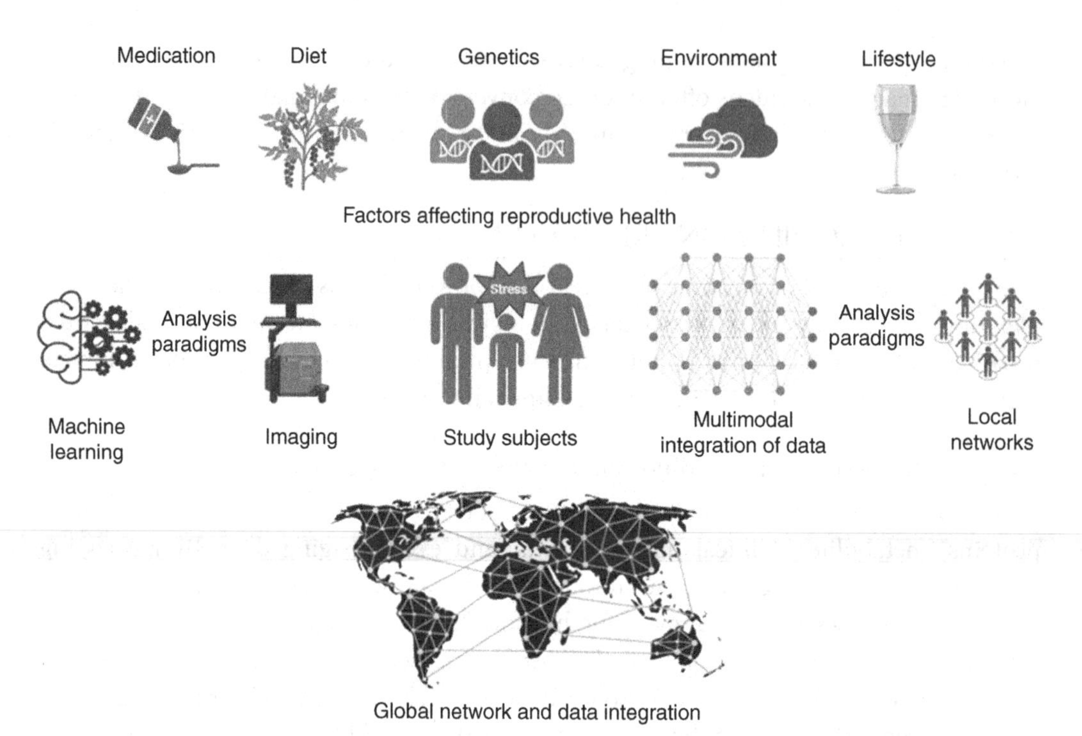

FIGURE 2.2 Factors affecting reproductive health and current strategies to develop a global framework of reproductive health information integration (Generated by using Biorender.com).

more integrated and nuanced understanding of pregnancy. This shift promises to revolutionize prenatal care and unlock a new era of personalized medicine for mothers and their babies. Factors affecting reproductive health and current global strategies are depicted in Figure 2.2.

2.7 CASE STUDY: PRETERM BIRTH AND ITS GLOBAL BURDEN

Despite concerted efforts and improved data availability, the global rate of PTB remained largely unchanged from 2010 to 2020. Ohuma et al. 2023 analyzed data from 103 countries and areas, revealing that an estimated 13.4 million babies were born preterm in 2020, representing 9.9% of all births. While this marks a slight decrease from 2010, the annual rate of decline was a meager −0.14%. Southern Asia and sub-Saharan Africa, collectively comprising 56% of live births, shouldered 65% of the global PTB burden in 2020. Worryingly, these regions also grapple with the most significant gaps in data quality, with none of the top 33 data quality countries residing in either region. Approximately 15% of all PTBs occurred at less than 32 weeks in 2020, highlighting the growing need for specialized neonatal care. Births at less than 28 weeks, despite representing only 4.2% of all PTBs, still numbered 567,800 globally. The fight against PTB demands renewed focus on equitable data collection, targeted interventions, and data-driven accountability. Only through these combined efforts can we truly bend the curve and improve the lives of millions of children and families affected by this global health challenge (Ohuma et al. 2023).

2.8 ROLE OF GUT MICROBIOME IN UNDERSTANDING REPRODUCTIVE HEALTH AND ITS IMPACTS

A healthy gut microbiome is essential for maintaining hormonal balance, which is key for successful conception. The gut microbiota interacts with your body's endocrine system, helping to regulate the production and metabolism of hormones such as estrogen and progesterone. Imbalances in these hormones can disrupt the menstrual cycle and make it difficult to conceive. Our gut microbiome also plays a role in nutrient absorption. Without proper nutrient absorption, your body may lack the essential vitamins and minerals needed for reproductive health. This can impact the quality of eggs and sperm, making it more challenging to conceive. To nurture your gut health and enhance fertility, it is important to focus on nourishing your gut microbiome. This can be achieved through a gut-friendly diet that includes specific nutrients and beneficial bacteria. Specific foods that can support your reproductive journey (Qi et al. 2021) are as follows:

A. **Fiber-rich fruits and vegetables:** Vegetables and fruits contain fiber, crucial for a healthy gut. Fiber aids in regular bowel movements and supports a diverse gut microbiome. Incorporate a range of colorful fruits and vegetables into your diet, such as berries, leafy greens, and cruciferous vegetables like broccoli and cauliflower (Makki et al. 2018).
B. **Incorporate probiotic-rich foods:** Probiotics, which are beneficial bacteria that help restore and maintain a healthy gut microbiota, are present in fermented foods such as yogurt, kefir, sauerkraut, and kimchi. These foods serve as natural reservoirs of probiotics, supporting a thriving gut microbiome (Irokanulo and Akalegbere 2019).
C. **Integrate omega-3 fatty acids:** Omega-3 fatty acids possess anti-inflammatory properties, promoting overall gut health. Enhance your diet with sources of omega-3 fatty acids, such as fatty fish (salmon, mackerel, sardines), chia seeds, and flaxseeds, to provide compounds that support fertility (Costantini et al. 2017).
D. **Wholegrains:** Wholegrains, such as oats, quinoa, and brown rice, are rich in fiber and can promote a healthy gut microbiome. Wholegrains are one of the most important carbohydrates for fertility as they are packed with nutrients and antioxidants (Seal et al. 2021).
E. **Plant proteins:** Lean proteins, such as chicken, turkey, fish, and legumes, provide important nutrients for fertility, including iron and zinc. These nutrients are essential for hormone production and support a healthy reproductive system. While animal protein can form part of a fertility diet for the amino acids, iron, and nutrients that may be harder to source from plant-based foods alone, opt for several "meat-free" meals across the week to increase your intake of plant-based protein. If you are getting excess protein from animal sources, this consumption usually comes with an increased saturated fat intake and a reduction in fiber intake (Lang et al. 2018).

Figure 2.3 depicts the taxonomic profiles in nonpregnant and pregnant women.

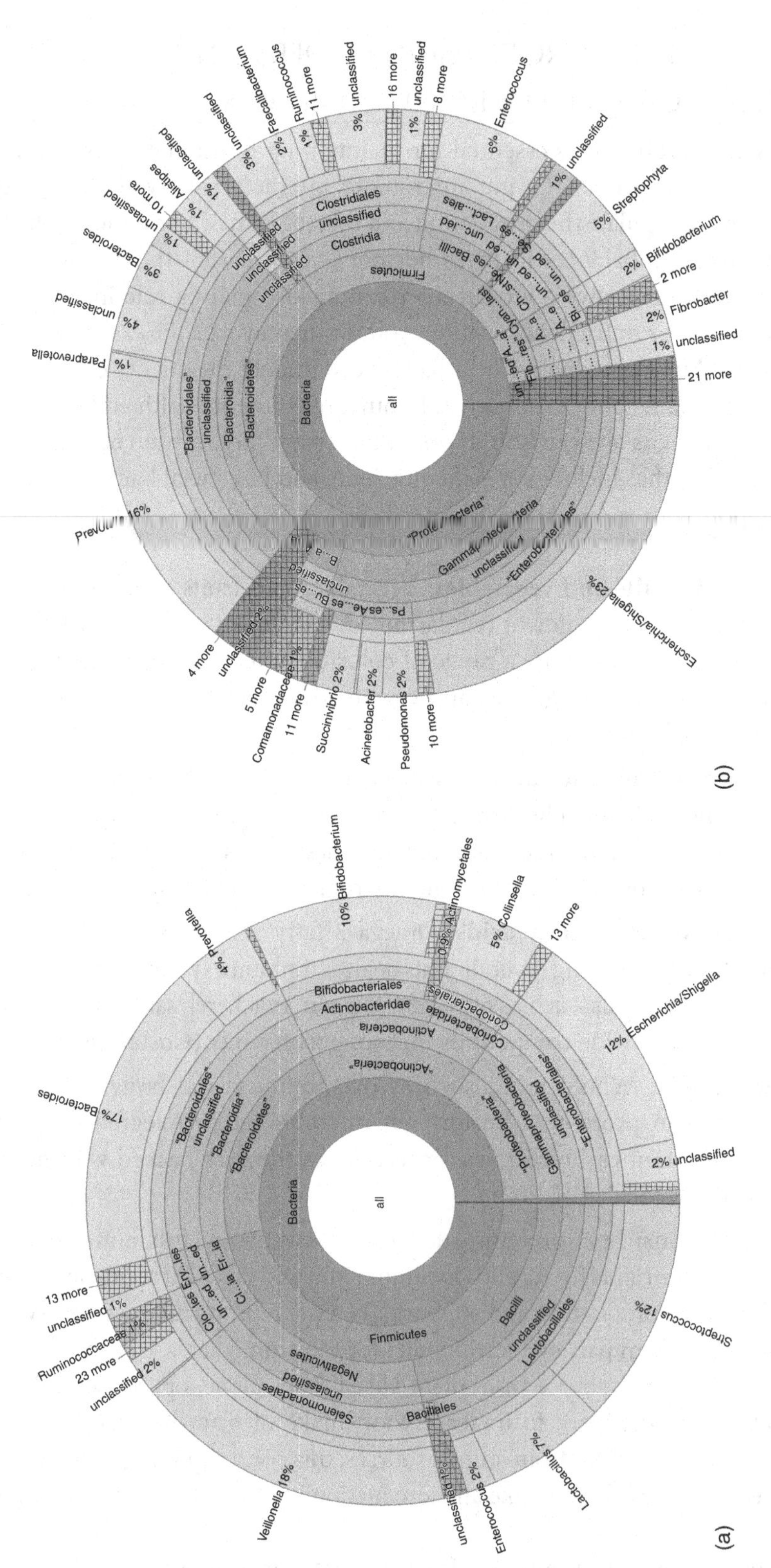

FIGURE 2.3 **(a)** Representative taxonomic elucidation of a non-pregnant woman; **(b)** representative taxonomic elucidation of a pregnant woman.

2.9 ONLINE DATA INTEGRATION STRATEGIES

- **Integrating Omics Data for Disease Gene Discovery**

Traditional methods for combining omics data in disease gene discovery can miss important leads. We propose a new framework based on desirability functions that seamlessly integrates diverse omics data and prioritizes candidate genes supported by strong evidence across multiple sources. This method is particularly suitable for analyzing diseases with limited and varied omics data, like sPTB. Using our integRATE R package, we successfully applied this framework to 10 sPTB studies, identifying both known and novel genes potentially involved in this condition (https://github.com/haleyeidem/integRATE).

- **The FP/RH Indicators Database: Monitoring Reproductive Health Programs**

Established in 2002, the Family Planning and Reproductive Health (FP/RH) Indicators Database (Figure 2.4) serves as a comprehensive resource for monitoring progress and evaluating programs in the area of reproductive health. It originated in response to two major trends: the 1994 International Conference on Population and Development (ICPD) which advocated for a broader definition of RH beyond family planning, and the increasing demand for accountability from both governments and donors involved in FP/RH initiatives. The database covers key areas such as safe motherhood, sexually

FIGURE 2.4 Screenshot of the FP/RH database homepage (www.data4impactproject.org/).

transmitted infections (STIs)/human immunodeficiency virus (HIV)/acquired immunodeficiency syndrome (AIDS), and women's nutrition, reflecting the evolving scope of RH beyond its traditional focus on family planning (Data for Impact Project 2024).

2.10 CONCLUSION

As the world faces rapid urbanization and decreasing social interactions, the understanding of reproductive health as a concept is of paramount importance for the global community. How we deal with complications for both the male and female subjects will be important for the sustenance of our species in the years to come and also prevent stigmatization in complex societal interactions and niche families. As a community, we need to find answers to the following pertinent questions:

1. Can AI unlock the power of combined biological, clinical, and social data to predict pregnancy complications?
2. Can readily available data replace expensive alternatives for accurate prediction?
3. Can machine learning unveil hidden connections to improve our understanding of adverse outcomes?

We believe that the answers to the above questions should help us in unraveling and spreading awareness regarding this aspect of human social interactions.

REFERENCES

Baharlou H, Canete NP, Cunningham AL, et al (2019) Mass cytometry imaging for the study of human diseases—applications and data analysis strategies. *Front Immunol* 10:2657. https://doi.org/10.3389/fimmu.2019.02657

Burton RJ, Albur M, Eberl M, Cuff SM (2019) Using artificial intelligence to reduce diagnostic workload without compromising detection of urinary tract infections. *BMC Med Inform Decis Mak* 19:171. https://doi.org/10.1186/s12911-019-0878-9

Carroll JE, Rentscher KE, Cole SW, et al (2020) Sleep disturbances and inflammatory gene expression among pregnant women: differential responses by race. *Brain Behav Immun* 88:654–660. https://doi.org/10.1016/j.bbi.2020.04.065

Clay K, Portnykh M, Severnini E (2018) *Toxic Truth: Lead and Fertility*. National Bureau of Economic Research, Cambridge, MA.

Cobb C, Kelly P, Williams K, et al (2017) The oral microbiome and adverse pregnancy outcomes. *IJWH* 9:551–559. https://doi.org/10.2147/IJWH.S142730

Costantini L, Molinari R, Farinon B, Merendino N (2017) Impact of omega-3 fatty acids on the gut microbiota. *IJMS* 18:2645. https://doi.org/10.3390/ijms18122645

Culos A, Tsai AS, Stanley N, et al (2020) Integration of mechanistic immunological knowledge into a machine learning pipeline improves predictions. *Nat Mach Intell* 2:619–628. https://doi.org/10.1038/s42256-020-00232-8

Cunningham JP, Ghahramani Z (2015) Linear dimensionality reduction: survey, insights, and generalizations. Journal Mach Learn Res 16:2859–2900. https://doi.org/10.48550/ARXIV.1406.0873

Handelman SK, Romero R, Tarca AL, et al (2019) The plasma metabolome of women in early pregnancy differs from that of non-pregnant women. *PLoS One* 14:e0224682. https://doi.org/10.1371/journal.pone.0224682

Harutyunyan H, Khachatrian H, Kale DC, et al (2019) Multitask learning and benchmarking with clinical time series data. *Sci Data* 6:96. https://doi.org/10.1038/s41597-019-0103-9

Huusko JM, Karjalainen MK, Graham BE, et al (2018) Whole exome sequencing reveals HSPA1L as a genetic risk factor for spontaneous preterm birth. *PLoS Genet* 14:e1007394. https://doi.org/10.1371/journal.pgen.1007394

Irokanulo E, Akalegbere MA (2019) Probiotics for gastrointestinal health and general wellbeing. *Intl J Probiot Prebiot* 15:22–29. https://doi.org/10.37290/ijpp2641-7197.15:22-29

Ismail Fawaz H, Forestier G, Weber J, et al (2019) Deep learning for time series classification: a review. *Data Min Knowl Disc* 33:917–963. https://doi.org/10.1007/s10618-019-00619-1

Kwak Y-T, Muralimanoharan S, Gogate AA, Mendelson CR (2019) Human trophoblast differentiation is associated with profound gene regulatory and epigenetic changes. *Endocrinology* 160:2189–2203. https://doi.org/10.1210/en.2019-00144

Lang JM, Pan C, Cantor RM, et al (2018) Impact of individual traits, saturated fat, and protein source on the gut microbiome. *mBio* 9:e01604-18. https://doi.org/10.1128/mBio.01604-18

Li E, Luo T, Wang Y (2019) Identification of diagnostic biomarkers in patients with gestational diabetes mellitus based on transcriptome gene expression and methylation correlation analysis. *Reprod Biol Endocrinol* 17:112. https://doi.org/10.1186/s12958-019-0556-x

Li L, Gong Y, Yin H, Gong D (2015) Different types of peptide detected by mass spectrometry among fresh silk and archaeological silk remains for distinguishing modern contamination. *PLoS One* 10:e0132827. https://doi.org/10.1371/journal.pone.0132827

Liu S, Wang Y, Yang X, et al (2019) Deep learning in medical ultrasound analysis: a review. *Engineering* 5:261–275. https://doi.org/10.1016/j.eng.2018.11.020

Makki K, Deehan EC, Walter J, Bäckhed F (2018) The impact of dietary fiber on gut microbiota in host health and disease. *Cell Host Microbe* 23:705–715. https://doi.org/10.1016/j.chom.2018.05.012

Marić I, Tsur A, Aghaeepour N, et al (2020) Early prediction of preeclampsia via machine learning. *Am J Obstet Gynecol MFM* 2:100100. https://doi.org/10.1016/j.ajogmf.2020.100100

Marzbanrad F, Stroux L, Clifford GD (2018) Cardiotocography and beyond: a review of one-dimensional Doppler ultrasound application in fetal monitoring. *Physiol Meas* 39:08TR01. https://doi.org/10.1088/1361-6579/aad4d1

Mei C, Yang W, Wei X, et al (2019) The unique microbiome and innate immunity during pregnancy. *Front Immunol* 10:2886. https://doi.org/10.3389/fimmu.2019.02886

Nweke HF, Teh YW, Al-garadi MA, Alo UR (2018) Deep learning algorithms for human activity recognition using mobile and wearable sensor networks: state of the art and research challenges. *Expert Syst Appl* 105:233–261. https://doi.org/10.1016/j.eswa.2018.03.056

Ohuma EO, Moller A-B, Bradley E, et al (2023) National, regional, and global estimates of preterm birth in 2020, with trends from 2010: a systematic analysis. *Lancet* 402:1261–1271. https://doi.org/10.1016/S0140-6736(23)00878-4

Park HJ, Cho HY, Cha DH (2021) The amniotic fluid cell-free transcriptome provides novel information about fetal development and placental cellular dynamics. *IJMS* 22:2612. https://doi.org/10.3390/ijms22052612

Qi X, Yun C, Pang Y, Qiao J (2021) The impact of the gut microbiota on the reproductive and metabolic endocrine system. *Gut Microbes* 13:1894070. https://doi.org/10.1080/19490976.2021.1894070

Saeys Y, Inza I, Larrañaga P (2007) A review of feature selection techniques in bioinformatics. *Bioinformatics* 23:2507–2517. https://doi.org/10.1093/bioinformatics/btm344

Santos-Lima CD, Mourão DDS, Carvalho CFD, et al (2020) Neuropsychological effects of mercury exposure in children and adolescents of the Amazon region, Brazil. *NeuroToxicology* 79:48–57. https://doi.org/10.1016/j.neuro.2020.04.004

Seal CJ, Courtin CM, Venema K, De Vries J (2021) Health benefits of whole grain: effects on dietary carbohydrate quality, the gut microbiome, and consequences of processing. *Comp Rev Food Sci Food Safe* 20:2742–2768. https://doi.org/10.1111/1541-4337.12728

Signorini MG, Pini N, Malovini A, et al (2020) Integrating machine learning techniques and physiology-based heart rate features for antepartum fetal monitoring. *Comput Methods Programs Biomed* 185:105015. https://doi.org/10.1016/j.cmpb.2019.105015

Souza RT, Cecatti JG, Mayrink J, et al (2019) Identification of earlier predictors of pregnancy complications through wearable technologies in a Brazilian multicentre cohort: Maternal Actigraphy Exploratory Study I (MAES-I) study protocol. *BMJ Open* 9:e023101. https://doi.org/10.1136/bmjopen-2018-023101

Tanacan A, Fadiloglu E, Beksac MS (2019) The importance of proteinuria in preeclampsia and its predictive role in maternal and neonatal outcomes. *Hypertens Pregnancy* 38:111–118. https://doi.org/10.1080/10641955.2019.1590718

Titus L, Hatch EE, Drake KM, et al (2019) Reproductive and hormone-related outcomes in women whose mothers were exposed in utero to diethylstilbestrol (DES): a report from the US National Cancer Institute DES Third Generation Study. *Reprod Toxicol* 84:32–38. https://doi.org/10.1016/j.reprotox.2018.12.008

United States Agency for International Development (USAID) (2024) Data for Impact 2024. In: Data for Impact 2024. www.data4impactproject.org/prh/overview. Accessed 30 Apr 2024.

Wainberg M, Merico D, Delong A, Frey BJ (2018) Deep learning in biomedicine. *Nat Biotechnol* 36:829–838. https://doi.org/10.1038/nbt.4233

Wang, D. (2001), Unsupervised learning: Foundations of neural computation. *AI Magazine*, 22:101–102. https://doi.org/10.1609/aimag.v22i2.1565

Yang H, Guo R, Li S, et al (2020) Systematic analysis of gut microbiota in pregnant women and its correlations with individual heterogeneity. *npj Biofilms Microbiomes* 6:32. https://doi.org/10.1038/s41522-020-00142-y

CHAPTER 3

Genomics and Transcriptomics in Reproductive Health

Sudeepti Kulshrestha, Payal Gupta, Nikhil H., Arunima N. P., Gopika S. Nair, Lalitha A., Shraddha Gurha, Vijay Pratap Singh, Mandar Bedse, Abhishek Sengupta, Pushpendra Singh, and Prashanth Suravajhala

3.1 INTRODUCTION

Reproductive health refers to a condition of optimal physical, mental, and social well-being, including all aspects of the reproductive system and its functions; and goes beyond the simple absence of disease or disability (Health et al. 1997). Reproductive health encompasses the state of both male and female reproductive systems throughout all stages of life (National Institute of Environmental Health and Sciences 2024) and entails the ability to engage in a safe sexual life, as well as the capacity to conceive and the independence to make choices regarding reproduction, including the timing and frequency of such decisions. Reproductive health is gaining recognition as a fundamental aspect of overall health and a significant factor in determining and measuring human social progress. General reproductive health is closely tied to overall well-being, as it serves as an indicator of health throughout childhood and adolescence and establishes the foundation for long-term health and life expectancy beyond the reproductive years. It is influenced by various health factors such as diet and environment, as well as by low birth weight, neonatal mortality, and morbidity (Mbizvo 1996). Poor reproductive health may lead to a range of reproductive disorders, causing substantial consequences for both males and females. Common female reproductive disorders include endometriosis, infertility or reduced fertility (difficulties in achieving pregnancy), menstrual disorders such as excessive or irregular bleeding, polycystic ovarian syndrome, pregnancy complications, sexually transmitted infections (STIs), uterine fibroids, and acquired immunodeficiency

DOI: 10.1201/9781003487548-3

syndrome (AIDS)/human immunodeficiency virus (HIV). Similarly, male reproductive disorders encompass various conditions, including impotence or erectile dysfunction, reduced sperm count, prostate cancer, STIs, and HIV/AIDS (National Institute of Environmental Health and Sciences 2024). Diagnosing and treating reproductive diseases may be challenging, which increases the likelihood of developing comorbidities. Furthermore, the persistent absence of research on reproductive health implies that the healthcare burden resulting from reproductive disorders, both in terms of immediate- and long-term care, is expected to keep rising (Mercuri and Cox 2022). The importance of prioritizing reproductive health cannot be overstated, as it plays a crucial role in shaping the current and future state of civilizations globally.

Omics technologies, including genomes, transcriptomics, metabolomics, and proteomics, are valuable tools in reproductive health research. They provide detailed information on the molecular pathways involved in reproductive processes and disorders. These techniques facilitate the identification of biomarkers that may be used for the timely detection, diagnosis, and prognosis of reproductive disorders. Additionally, they provide a comprehensive comprehension of reproductive health, hence facilitating the implementation of tailored medical strategies. Moreover, they possess the capacity to increase clinical results, direct therapeutic actions, and improve procedures in reproductive healthcare.

This chapter examines the contributions of genomics and transcriptomics to the field of reproductive health. Genomics is the study of genomes, focusing on their structure, function, and evolution. It involves the use of messenger molecules and enzymes to control the synthesis of proteins. However, transcriptomics is an approach that thoroughly examines the messenger RNA (mRNA) molecules present in a cell, tissue, or organism. It offers detailed information on the total number of mRNA molecules and the specific concentration of each RNA molecule (Kumari et al. 2024). Integrating genomics and transcriptomics technologies with reproductive health enables advances in diagnoses, treatments, and personalized medicine.

3.2 IMPORTANCE OF OPTIMAL REPRODUCTIVE HEALTH

The importance of reproductive health lies in empowering women to make well-informed choices about their bodies and overall well-being. This includes education about the physiological changes of the reproductive system (from adolescence till menopause), strategies for family planning, methods of contraception, prenatal and postnatal healthcare, and testing for reproductive disorders. Access to reproductive healthcare could reduce maternal mortality rates as adequate medical care can avert pregnancy-related complications. Furthermore, reproductive health plays a crucial role in fostering gender equality by encouraging women to choose education and economic prospects over early parenthood and childcare responsibilities. Additionally, it serves as a barrier against the spread of sexually transmitted illnesses and yields beneficial outcomes for future generations, resulting in more prosperous communities, more resilient families, and improved well-being for mothers and children. Comprehending the intricacies of the female reproductive system, which encompasses the menstrual cycle, fertility, conception, pregnancy, and more, is crucial for maintaining optimal reproductive well-being.

Similarly, the significance of men's reproductive health extends to its impact on fertility, sexual function, and general well-being. Male infertility and sexual dysfunction may cause significant stress and strain in relationships. Furthermore, research has shown a connection between male infertility and chronic illnesses such as cardiovascular disease and diabetes (Choy and Eisenberg 2018). Factors such as obesity, smoking, and exposure to environmental toxins can adversely affect sperm quality and reproductive function, serving as early indicators of potential health risks (Leisegang et al. 2021, Choy and Eisenberg 2018). By prioritizing reproductive health, men not only enhance their prospects of fathering healthy offspring but also mitigate the potential for developing long-term health problems in the future.

Thus, it is important to advocate for and maintain ideal reproductive health in both males and females, as it is essential for personal welfare and tackling wider public health issues.

3.3 NEXT-GENERATION SEQUENCING TECHNOLOGIES IN REPRODUCTIVE HEALTH

Next-generation sequencing or NGS is a high-throughput technique capable of rapidly sequencing whole genomes or targeted sections of DNA or RNA. NGS can profoundly transform several domains of biological research, including disease diagnosis, prognosis, and the determination of therapeutic interventions. NGS expedites the identification of genetic variations linked to diseases, facilitating the implementation of personalized medicine strategies and precision medicines. Furthermore, it simplifies the investigation of transcriptomics, uncovering complex patterns of gene expression and regulatory processes.

3.3.1 Genomics-Based Approaches

Figure 3.1 demonstrates the general workflow followed in genomic studies.

3.3.1.1 Whole-Genome Sequencing

Whole-genome sequencing, often known as WGS, is a technique used to ascertain the whole DNA sequence of an individual's genome. It enables a comprehensive analysis of genomic variations, including single-nucleotide polymorphisms or SNPs, insertions, deletions, and structural variants. WGS is increasingly being used as the preferred technique for diagnosing rare and unknown diseases at the molecular genetic level, as well as for identifying biomarkers that may be targeted for treatment. When compared to other molecular genetic approaches, WGS can detect a wide range of genomic variations and removes the necessity for sequentially conducting genetic testing (Bagger et al. 2024). WGS can be used to detect genetic predispositions to reproductive disorders, including infertility, chromosomal abnormalities, and genetic syndromes that impact fertility.

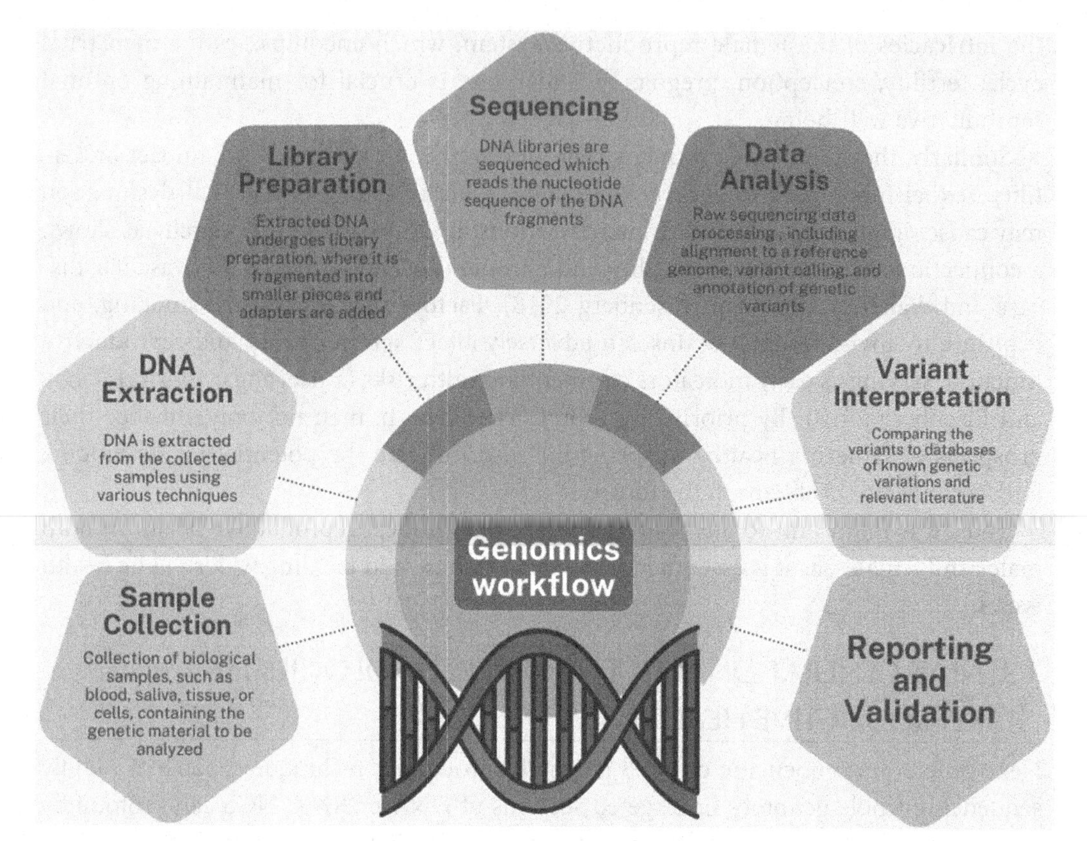

FIGURE 3.1 Workflow followed in genomic studies.

3.3.1.2 Whole-Exome Sequencing

The exome refers to the portion of the genome that consists of all the exons of protein-coding genes. It makes up around 1–2% of the genome, with the exact percentage varying depending on the species (Warr et al., 2015). Approximately 85% of mutations that cause diseases are found in the exome (Seaby et al. 2016). As a result, whole-exome sequencing (WES) has gained popularity as a more affordable alternative to WGS. WES offers a balance between cost, coverage of the genome, diagnostic success, and interpretability. WES can help to uncover genetic changes that are linked to fertility, infertility, and reproductive disorders. By prioritizing coding sequences, it provides valuable information on uncommon genetic variations, Mendelian disorders, and prospective targets for therapeutic interventions. This contributes to the progress of tailored strategies in reproductive medicine and genetic counseling.

3.3.1.3 Epigenomics

Epigenomics is the approach used to investigate the chemical alterations of DNA that have a significant impact on the expression and regulation of gene function. Epigenomic regulation is a complex and dynamic process wherein genomic regulation is influenced by external influences, such as diet, nutrition, and environmental factors. These heritable alterations do not change the genetic sequence but play a crucial role in determining the final characteristics and are necessary for various stages of development (Martin-Sanchez et al. 2014). Epigenomics in reproductive health can be used to explore heritable changes in gene expression, offering insights into how environmental variables affect fertility, pregnancy outcomes, and the well-being of children. Epigenomics reveals complex molecular pathways underlying reproductive problems by studying DNA methylation, histone changes, and noncoding RNA regulation, thus providing potential opportunities for targeted therapies and preventive measures.

3.3.1.4 Genome-Wide Association Studies

Genome-wide association studies, also known as GWAS, seek to identify associations among phenotypes and genotypes by analyzing differences in the frequency of genetic variants across individuals with similar ancestry but differing phenotypic traits. GWAS may include copy-number variants (CNVs) or SNPs in the human genome. GWAS studies often detect clusters of linked SNPs that have a statistically significant association with the particular trait of interest. These clusters are referred to as genomic risk loci (Uffelmann et al. 2021). GWAS in reproductive health may be used to investigate the genetic underpinnings of fertility, reproductive diseases, and pregnancy outcomes by examining large groups of individuals to pinpoint genetic variations linked to these characteristics. By conducting thorough genome-wide scans, GWAS identifies specific candidate genes and genetic regions that are associated with reproductive traits. This information is valuable for clinical diagnoses, risk evaluation, and individualized treatment approaches in the field of reproductive medicine.

3.3.1.5 Microbiomics

The human body contains microbiomes in several ecological niches, such as the skin surface, mouth cavity, esophagus, lungs, digestive system, and other regions. The microbiota is comprised of bacteria, viruses, fungi, archaea, and phages with bacteria being the most abundant in terms of species abundance. The use of 16S rRNA and WGS techniques has facilitated the detection and categorization of multiple distinct operational taxonomic units (OTUs), some of which may be cultured and identified, while others remain unidentified. Microbial dysbiosis is caused by alterations in the makeup of microbiome species or changes in the abundance of certain species, leading to an imbalance in the microbiome, and giving rise to disease states and conditions. It may also occur when some bacterial species in a microbiome acquire qualities that make them more virulent and pathogenic, and these harmful characteristics then alter the way the host organism is affected in its unique environment (Manos 2022).

The markers often used to analyze microbial communities in microbiome investigations include 16S rRNA, 18S rRNA, and ITS rRNA. These markers allow for the characterization of bacterial, archaeal, and fungal diversity, respectively.

These diverse genomics technologies may be used to address common reproductive diseases and disorders that impact both males and females.

3.3.1.6 Application of Genomics in Female Reproductive Health

Genotyping of 400 women with polycystic ovary syndrome or PCOS and 480 healthy women identified several gene variations, including *rs6166* of the *FSHR* gene, *rs13405728* of the *LHCGR* gene, *rs13429458* of the *THADA* gene, and others, which may contribute to the hereditary risk of PCOS in Asian women (Wan et al. 2021). Furthermore, GWAS demonstrated that the genes *ERBB4*, *YAP1*, and *WWTR1* have a role in the complex etiology of PCOS. This is achieved by the activation of epidermal growth factor receptors or EGFRs, and the Hippo pathway, as reported by Zhang et al. in 2020. Moreover, the genotyping analysis of 152 women suffering from primary dysmenorrhea versus 150 individuals without any health issues revealed that there were substantial statistical differences in the *ESR1* gene *PvuII* polymorphism between the patients and controls (Ozsoy et al. 2016).

In addition, recent epigenome-wide association studies (EWAS) have shown a strong link between factors like maternal smoking during pregnancy, maternal folate levels, air pollution caused by traffic, exposure to particulate matter during pregnancy, and significant alterations in methylation patterns in newborns. The presence of these changes in methylation has been linked to a higher likelihood of preterm delivery (Park et al. 2020). Similarly, WGS, WES, and trio bioinformatics analysis were conducted on eight couples who experienced unexplained recurrent miscarriage (URM), which revealed six candidate genes that were previously unknown, namely *ATP2A2, SSPO, NAP1L1, PLXNB2, NRK*, and *RYR2*. These genes were found to be associated with unexplained first-trimester euploid miscarriages (Wang et al. 2023).

Furthermore, WGS has been used to investigate Gonorrhea, a sexually transmitted illness caused by the *Neisseria* bacteria. The analysis of WGS data has played a significant role in detecting outbreaks of *Neisseria meningitidis*, characterizing isolates of both *Neisseria gonorrhoeae* and *N. meningitidis*, and enhancing disease monitoring. In addition, it has facilitated the integration of pathogen and patient genes sequence data to enhance our comprehension of patient response and vulnerability to gonorrhea infection (Harrison et al. 2017).

An investigation of the 39 16S rRNA vaginal microbiome sequences categorized into preterm birth and term birth, showed that the presence of beneficial microbe like *Lactobacillus*, as well as pathogenic microbes like *Gardnerella*, *Prevotella*, and *Atopobium*, and particular genes such as *FANCM*, *PTXA*, and *GPX*, are linked to the likelihood of preterm birth (Kulshreshtha et al. 2023). It has also been observed that the vaginal microbiome and its composition are extremely dynamic and vary significantly across women of various races, ethnicities, and sampling sites (Kulshrestha et al. 2024). Additionally, the gut microbiome of 20 women with PCOS and 20 controls, revealed that gamma-aminobutyric acid (GABA)-producing bacteria such as *Bacteroides fragilis*, *Parabacteroides distasonis*, and *Escherichia coli* have been shown to have a positive association with LH:FSH (luteinizing hormone to follicle-stimulating hormone) ratios and serum LH levels, hence playing a role in triggering PCOS (Liang et al. 2021).

3.3.1.7 Application of Genomics in Male Reproductive Health

Array-based comparative genomic hybridization, also known as aCGH, was performed on a group of 37 men who had meiotic arrest, 194 men who exhibited the Sertoli cell-only phenotype, and 21 control men, along with exome data from 2,030 men determined the reason for infertility has been to be heterozygous and homozygous deletions of the *SYCE1* gene, along with heterozygous deletions affecting the *EIF2B2, MLH3, TEKT5, CLPP*, and *SLX4* genes (Wyrwoll et al. 2022).

A consanguineous Turkish family with idiopathic nonobstructive azoospermia was studied utilizing WES analysis to determine the genetic etiology of infertility. The investigation identified a specific genetic variation in the *GTF2H3* gene on chromosome 12. This variation was found as a homozygous variant in patients with non-obstructive azoospermia and as a heterozygous variant in patients with oligospermia. It was theorized that these variations might have a detrimental effect on the signaling of vitamin A, crucial for spermatogenesis (Clavijo et al. 2018).

Epigenomic research has also indicated that epigenetic alterations and factors play a role in regulating the activity of genes, including *Fgf8*, that are crucial for the growth and operation of gonadotropin-releasing hormone (GnRH) neurons. These factors also contribute to the development of Kallmann syndrome and congenital hypogonadotropic hypogonadism, conditions characterized by the inability of the testes to produce androgens and sperm (Linscott and Chung 2020). Additionally, an EWAS was conducted on DNA methylation patterns related to HIV infection. The study included 261 patients infected with HIV and 117 healthy individuals. The results revealed that 20 *CpG* sites showed a significant association with HIV infection. Among these sites, 14 were found to have lower levels of methylation (hypomethylated) in HIV-infected individuals, while 6 sites showed higher levels of methylation (hypermethylated) (Arumugam et al. 2021).

3.3.2 Transcriptomics-Based Approaches

The genetic information of an organism is stored in its genome's DNA and is manifested via transcription (Lowe et al. 2017) refers to the comprehensive study of all RNA transcripts present in a cell or tissue at a certain moment, offering valuable information on the patterns of gene expression, regulation, and functional variability. It includes techniques like RNA sequencing (RNA-Seq) to quantify and evaluate several types of RNA molecules, including mRNA, noncoding RNA, and splice variants. It provides a detailed view of the transcriptome's changing landscape in different biological situations. Transcriptomics profiling can be used in reproductive health to analyze the whole collection of RNA transcripts in cells, tissues, or species to gain insights into the patterns of gene expression and the regulatory networks that play a role in fertility, reproduction, and associated illnesses. The general workflow followed in transcriptomics studies is depicted in Figure 3.2.

3.3.2.1 Microarray Sequencing

Microarray transcriptome analysis entails quantifying the levels of certain transcripts by hybridizing the m to a set of complementary probes (Lowe et al. 2017). A microarray

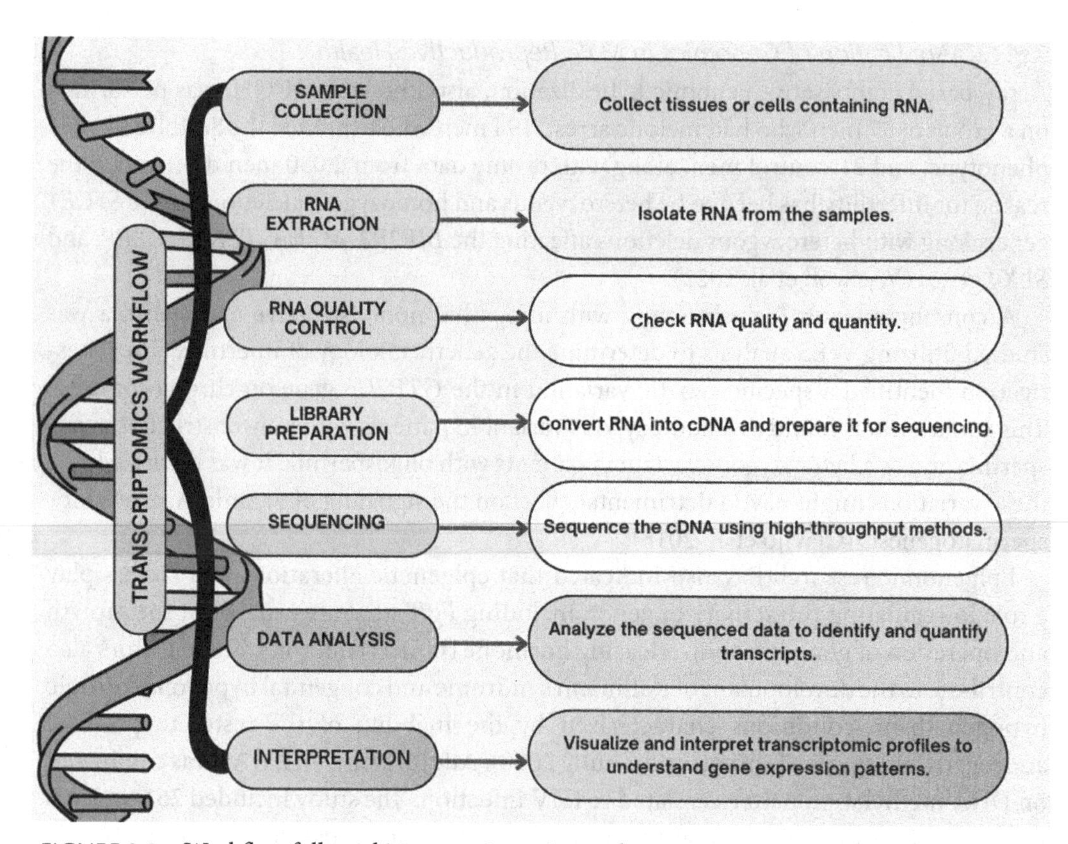

FIGURE 3.2 Workflow followed in transcriptomics studies.

begins with a collection of brief oligonucleotide probes that represent genomic DNA. Transcripts are obtained from cell or tissue samples to be examined, marked with fluorescent dyes, combined with arrays, cleansed, and analyzed using a laser. Probes that are complementary to transcribed RNA bind to their corresponding target by hybridization. Transcripts are marked with fluorescent dyes, allowing light intensity to serve as an indicator of gene expression (Malone and Oliver 2011). Affymetrix high-density arrays and spotted oligonucleotide arrays were the preferred methods for transcriptional profiling in the early 2000s. Microarray technology enables the simultaneous analysis of thousands of transcripts at a far lower cost per gene and with decreased labor requirements. The results from gene expression microarrays have yielded significant insights into the deployment of the transcriptome in various tissues and cell types, the changes in gene expression during different developmental stages and disease phenotypes, and the variations across and within species (Lowe et al. 2017).

3.3.2.2 Bulk-RNA Sequencing

Bulk-RNA Sequencing, often known as RNA-Seq, is a method used to sequence complementary DNAs (cDNAs) derived from transcripts, to determine their abundance based on the count of each transcript. RNA-Seq utilizes extensive transcriptome sampling using several short fragments from a transcriptome. This allows for the computational rebuilding of the primary transcript by matching the reads to one another via de novo assembly or to a reference genome. The RNA-Seq technique offers a substantial advantage over microarray transcriptomes due to its usual range of 5 orders of magnitude (Lowe et al. 2017). The main difference between RNA-Seq and microarrays is that RNA-Seq allows for comprehensive whole transcriptome sequencing, whereas microarrays only examine specified transcripts or genes (Rao et al., 2019).

3.3.2.3 Single-Cell RNA Sequencing

Most cells in humans have the same genetic makeup. Nevertheless, the transcriptome data inside each cell provides insight into the specific functioning of a particular group of genes. Traditional investigation of the transcriptome or epigenome utilizing RNA-Seq may only capture the collective signals from tissues or organs, without the ability to distinguish between specific cellular changes. Single-cell RNA sequencing (ScRNA-Seq) facilitates the investigation of the transcriptome at the cellular level, enabling the inspection of millions of cells in a single experiment. Single-cell isolation and capture is the process of extracting individual cells from tissue to collect precise genetic and biochemical information. This allows for the study of specific genetic and molecular processes. By analyzing the transcriptome of each cell, we can classify, define, and distinguish them, making it easier to identify rare cell groups that have functional importance (Jovic et al. 2022).

3.3.2.4 miRNA, ncRNA, LncRNA, and Circular RNA Profiling

Recent breakthroughs in molecular biology have revealed a diversity of noncoding RNA molecules (ncRNA), such as long noncoding RNAs or lncRNAs, microRNAs (miRNAs), and circular RNAs or circRNAs. These molecules play a crucial role in controlling gene expression at different stages and are also engaged in a wide range of physiological functions. ncRNA refer to RNA molecules that do not perform protein encoding. However, it is important to note that these RNAs still perform specific functions (Panni et al. 2020).

MicroRNAs are a set of small noncoding RNAs that regulate gene expression after transcription by specifically binding to mRNAs (Hussen et al. 2021). This binding often occurs at a short complementary sequence, frequently found in the 3′ untranslated region (UTR) of the mRNA. Similarly, lncRNAs are a diverse group of RNAs that may be transcribed from long intergenic noncoding RNAs (lncRNAs), or from any section of protein-coding genes, in either the same or opposite direction. LncRNAs are characterized by their length, which must exceed 200 nucleotides, and their inability to code for proteins. Additionally, circular RNAs are RNA molecules that have a single strand and are formed when the 3′ and 5′ ends are joined together, creating a circular shape. They are formed by a process known as back-splicing and typically consist of exons, introns, or noncoding intergenic regions (Panni et al. 2020).

The distinct expression patterns of noncoding RNAs in different illness conditions substantiate their potential as promising biomarkers, facilitators of disease-causing pathways, and possible targets for therapeutic interventions (Ali et al. 2021).

3.3.2.5 Spatial Transcriptomics

The term "spatial," originating from the Latin word spatium, which means "space," is used to characterize the relative positioning of things. The spatial idea is of great significance in biology as it enables us to depict biological interaction networks, where each constituent is impacted by its environment. For example, a thorough comprehension of the chemical characteristics of individual cells in multicellular organisms can only be achieved when we know their precise physical positions. On this note, the subcellular localization of the RNA molecule might provide significant insights into the transcriptional snapshot and its true biological significance (Asp et al. 2020). Thus, spatial transcriptomics is a technique that reveals the location-specific variation of gene expression patterns. It gives valuable insights into aspects that were previously unknown using ScRNA-seq techniques, which do not capture spatial information (Park et al. 2023).

The abovementioned transcriptomics strategies may be used to address common reproductive disorders and illnesses that impact both males and females.

3.3.2.6 Application of Transcriptomics in Female Reproductive Health

scRNA-Seq analysis has shown that increased levels or improved activity of the transcription factors *SREBF1* (which controls genes involved in cholesterol acquisition such as NPC1, *LIPA*, *LDLR*, and others) and *GATA6* (which regulates the expression of genes encoding steroidogenic enzymes like *CYP17A1*), *SP1*, and *NR5A2* are involved in the development of hyperandrogenemia associated with PCOS (Harris et al. 2023). In silico transcriptomics research identified several differentially expressed genes, including *CENPF, CEP55, DLGAP5, KIF20A*, and others, which have a significant impact on endometrial receptivity during the transition from the proliferative to the secretory phase (Gupta et al. 2023).

The analysis of RNA-Seq data from umbilical cord blood leukocytes identified 148 genes that were significantly differentially expressed. In particular, the preterm birth group exhibited a large increase in the expression of genes associated with cell cycle and metabolism, while showing a significant decrease in the expression of genes associated with immunological and inflammatory signaling, as compared to the full-term group. (Vora et al. 2019). Additionally, analysis of scRNA-Seq data from maternal macrophages in the chorioamniotic membranes identified distinct alterations in gene expression that were associated with premature labor. Furthermore, the maternal circulation in preterm labor showed modulation of several placental scRNA-Seq transcriptional signatures related to macrophages, monocytes, and activated T-cells as gestation progressed (Pique-Regi et al. 2019).

ScRNA-Seq analysis revealed distinct gene expression patterns between cells infected with HIV and those that were not infected. The presence of latency signatures in cells infected with HIV is linked to cellular exposure, responsiveness to T-cell receptor activation, and viral tropism (Zhang et al. 2023).

3.3.2.7 Application of Transcriptomics in Male Reproductive Health

Analysis using RNA-Seq and scRNA-Seq techniques revealed that decreased Wnt signaling and three specific transcription factors (*ARNTL, CTCF,* and *AR*) are involved in the development of nonobstructive azoospermia in males (Zeng et al. 2023). Similarly, a comparative analysis of scRNA-Seq data was conducted to compare males with Klinefelter syndrome, nonobstructive azoospermia, cryptozoospermia, and controls. The results revealed a unique gene expression pattern in microvascular-associated cells of males with Klinefelter syndrome. This pattern suggests an excessive activation of endothelial cells (ECs), disorganized formation of blood vessels, and the presence of immature vessels with compromised integrity (Johannsen et al. 2023). Additionally, transcriptome data analysis has connected differentially expressed lncRNAs and mRNAs with their target genes to the buildup of unfolded proteins in the sperm endoplasmic reticulum (ER), oxidative stress, sperm cell apoptosis, and ER stress induced by the PERK-EIF2 (protein kinase R-like ER kinase-eukaryotic initiation factor 2) pathway in individuals with oligozoospermia, thus affecting male fertility (Sun et al. 2021).

3.4 CLINICAL APPLICATIONS OF GENOMICS AND TRANSCRIPTOMICS IN REPRODUCTIVE HEALTH

3.4.1 Assisted Reproductive Technologies

Infertility is a medical disorder when a person is unable to have a successful pregnancy while participating in frequent, unprotected intercourse for a duration of 12 months or more. It may also refer to a person's inability, either alone or with their partner, to reproduce. The first step in diagnosing infertility in any patient is obtaining a comprehensive medical history, including details on their sexual and reproductive history, as well as any signs of decreased reproductive function that may be associated with advanced age or physical examination results (Practice Committee of the American Society for Reproductive Medicine 2020).

In recent years, assisted reproductive technologies (ART) have become a secure solution for couples experiencing fertility issues. ART approaches include controlled in vitro fertilization (IVF), vitrification, intracytoplasmic sperm injection (ICSI), and controlled ovarian stimulation (COS) (Sánchez-Pavón et al. 2022). Typically, the procedure often involves the use of donated gametes due to diagnosed or idiopathic infertility in the individuals seeking ART. Gamete donation refers to the act of donating either eggs (oocytes) or sperm to help another person or couple in their efforts to have a child. Women who have hypogonadotropic hypogonadism, advanced maternal age, reduced ovarian reserve, low oocyte/embryo quality, or have had many unsuccessful efforts to conceive after fertility treatments are recommended to consider using an oocyte donor. Likewise, the reasons for using a sperm donor include Azoospermia, severe oligospermia, or other notable abnormalities in sperm or seminal fluid; malfunction in ejaculation; or other serious infertility in males that has not improved with earlier ART treatments. Other indications include women/men who are known to be afflicted by or carriers of a major genetic abnormality, as well as those who have a family history of a heritable disease for which testing is currently unavailable. In addition, the existence of an incurable sexually transmitted illness or a situation where a male partner is Rh-positive and his female spouse is Rh-negative with a severe case of Rh-isoimmunization are additional factors that indicate the need for ART (Lilienthal and Cahr 2020).

Although ART offers several benefits, it also carries a significant risk of genetic problems, including trisomy, and illnesses linked to harmful changes in the mitochondrial DNA. Research shows a higher likelihood of placenta previa, preeclampsia, preterm birth, intrauterine growth restriction, and placental abruption, in pregnancies resulting from IVF. Furthermore, ART has been associated with several rare imprinting syndromes, in addition to individual congenital abnormalities, including Beckwith–Wiedemann, Angelman, Prader–Willi, and Silver–Russell syndromes. The higher occurrence of these illnesses after IVF indicates the involvement of improper regulation of epigenetic factors in the emergence of adverse outcomes after ART procedures (Mani et al. 2020). Infants conceived with ART also have a threefold higher likelihood of being delivered with low birth weight. The birth weight and preterm of ART-conceived neonates may be significantly influenced by several factors, including the choice between fresh or frozen embryo transfers, the number of embryos transferred, and the method of endometrial preparation (Reig and Seli 2019).

Genomics and transcriptomics techniques have revolutionized ART by uncovering the genetic variables that impact fertility and the growth of embryos. By using methods such as preimplantation genetic testing (PGT), they facilitate the identification and selection of viable embryos for transfer in the process of IVF. Transcriptomic analyses improve embryo selection strategies by identifying important gene expression patterns, increasing success rates, and assuring safer operations.

3.4.2 Preimplantation Genetic Testing

PGT is a medical treatment that entails extracting one or more nuclei from an egg or an embryo to examine for genetic alterations or aneuploidy before transferring the embryo (Lilienthal and Cahr 2020). PGT is founded on the idea that the genetic makeup of cells obtained from the preimplantation embryo accurately reflects the population of cells that will ultimately develop into the fetus (Herbert et al. 2023). The primary objective of PGT is to mitigate the likelihood of conceiving a child who may inherit a genetic condition or chromosomal abnormalities. The general steps followed in PGT have been depicted in Figure 3.3.

There are three distinct categories of PGT: PGT for aneuploidy (PGT-A), PGT for structural rearrangements (PGT-SR), and PGT for monogenic diseases (PGT-M) (Lilienthal and Cahr 2020).

The fundamental objective of PGT-A is to evaluate the presence or absence of aneuploidy, therefore minimizing the likelihood of implantation failure, miscarriage (Lilienthal and Cahr 2020), or the occurrence of congenital diseases associated with trisomy (such as Edwards syndrome and Down syndrome) (Herbert et al. 2023). PGT-SR enables the selection of embryos with a presumably balanced set of chromosomes (Herbert et al. 2023). Structural chromosomal rearrangements (SRs) result in infertility, recurrent failure of embryo implantation, miscarriages, and children with congenital abnormalities, even when the carrier parents seem to have no visible physical characteristics (Griffin and Ogur 2023). PGT-M is used to identify certain genetic abnormalities in embryos. It is

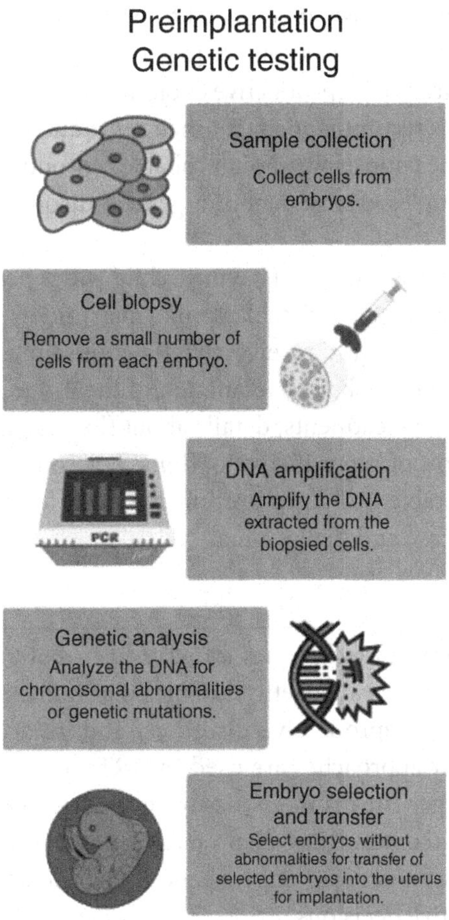

FIGURE 3.3 General steps followed for preimplantation genetic testing.

suited for persons who are afflicted by or have a genetic history of an autosomal dominant condition. It is also used for those who are at risk of having children with an autosomal recessive or X-linked disorder. Additional rationales for doing PGT-M include the purpose of human leukocyte antigen (HLA)-matching for a child requiring a stem cell transplant, as well as the selection of embryos that do not carry the danger of hemolytic illness of the infant for people with antigen sensitivity resulting from a previous pregnancy (Lilienthal and Cahr 2020).

Genomics and transcriptomics enhance PGT by precisely identifying chromosomal abnormalities and genetic disorders in embryos before implantation. This improves the selection process for transfer in assisted reproduction techniques such as IVF and reduces the chances of inheriting diseases.

3.4.3 Carrier Screening

Carrier screening refers to genetic testing conducted on individuals who do not exhibit any visible signs of a genetic condition but may have one variation allele within a gene or genes linked with a specific disease. It is crucial to perform carrier screening and counseling before conception as this allows couples to get knowledge about their reproductive risk and explore a wide variety of reproductive choices. If a person is identified as a carrier for a particular disease, it is recommended that their reproductive partner undergo testing to enable informed genetic counseling about probable reproductive consequences. If it is determined that both couples are carriers of a genetic disorder, they may be provided with genetic counseling.

For example, couples who are at risk of having a child with a hemoglobinopathy, such as sickle-cell anemia, α-thalassemia, or β-thalassemia, may find it advantageous to undergo carrier screening and subsequently receive genetic counseling. This counseling would involve an assessment of their risk, an explanation of the progression of these disorders, information about potential treatments, details about the availability of prenatal genetic testing, and an exploration of reproductive options. For some couples, the use of PGT with IVF may be a preferable option to prevent the need for terminating a pregnancy afflicted by sickle cell disease and thalassemia.

3.5 CONCLUSION

Genomics and transcriptomics techniques are critical in both female and male reproductive health because they reveal genetic variables that influence fertility, identify possible risks for infertility or reproductive disorders, and guide personalized treatment methods. These molecular approaches are used in ART to screen embryos for chromosomal abnormalities and genetic disorders, assuring the transfer of healthy embryos and enhancing pregnancy outcomes. They also enable complete carrier screening, which identifies people who are at risk of passing on genetic disorders to their kids, allowing for more informed reproductive decisions.

REFERENCES

Ali SA, Peffers MJ, Ormseth MJ, et al (2021) The non-coding RNA interactome in joint health and disease. *Nat Rev Rheumatol* 17:692–705. https://doi.org/10.1038/s41584-021-00687-y

Arumugam T, Ramphal U, Adimulam T, et al (2021) Deciphering DNA methylation in HIV infection. *Front Immunol* 12:795121. https://doi.org/10.3389/fimmu.2021.795121

Asp M, Bergenstråhle J, Lundeberg J (2020) Spatially resolved transcriptomes—next generation tools for tissue exploration. *BioEssays* 42:1900221. https://doi.org/10.1002/bies.201900221

Bagger FO, Borgwardt L, Jespersen AS, et al (2024) Whole genome sequencing in clinical practice. *BMC Med Genom* 17:39. https://doi.org/10.1186/s12920-024-01795-w

Choy JT, Eisenberg ML (2018) Male infertility as a window to health. *Fertil Steril* 110:810–814. https://doi.org/10.1016/j.fertnstert.2018.08.015

Clavijo RI, Arora H, Gibbs E, et al (2018) Whole exome sequencing of a consanguineous Turkish family identifies a mutation in GTF2H3 in brothers with spermatogenic failure. *Urology* 120:86–89. https://doi.org/10.1016/j.urology.2018.06.031

Griffin DK, Ogur C (2023) PGT-SR: a comprehensive overview and a requiem for the interchromosomal effect. *DNA* 3:41–64. https://doi.org/10.3390/dna3010004

Gupta P, Dube S, Priyadarshini P, et al (2023) Deciphering key genes of proliferative and secretory phase using integrated transcriptomics and network analysis. *Microbiol Biotechnol Lett* 51:317–324. https://doi.org/10.48022/mbl.2304.04008

Harris RA, McAllister JM, Strauss JF (2023) Single-cell RNA-Seq identifies pathways and genes contributing to the hyperandrogenemia associated with polycystic ovary syndrome. *IJMS* 24:10611. https://doi.org/10.3390/ijms241310611

Harrison OB, Schoen C, Retchless AC, et al (2017) *Neisseria* genomics: current status and future perspectives. *Pathog Dis* 75:ftx060. https://doi.org/10.1093/femspd/ftx060

National Research Council (US) Panel on Reproductive Health, Tsui AO, Wasserheit JN, Haaga JG (1997) Introduction. In: *Reproductive Health in Developing Countries: Expanding Dimensions, Building Solutions*. Washington, DC: National Academies Press (US).

Herbert M, Choudhary M, Zander-Fox D (2023) Assisted reproductive technologies at the nexus of fertility treatment and disease prevention. *Science* 380:164–167. https://doi.org/10.1126/science.adh0073

Hussen BM, Hidayat HJ, Salihi A, et al (2021) MicroRNA: a signature for cancer progression. *Biomed Pharmacother* 138:111528. https://doi.org/10.1016/j.biopha.2021.111528

Johannsen EB, Skakkebæk A, Kalucka JM, et al (2023) The testicular microvasculature in Klinefelter syndrome is immature with compromised integrity and characterized by excessive inflammatory cross-talk. *Hum Reprod* 38:2339–2349. https://doi.org/10.1093/humrep/dead224

Jovic D, Liang X, Zeng H, et al (2022) Single-cell RNA sequencing technologies and applications: a brief overview. *Clin Transl Med* 12:e694. https://doi.org/10.1002/ctm2.694

Kulshreshtha S, Narad P, Singh B, et al (2023) Identification of distinct vaginal microbiota signatures contributing toward preterm birth using an integrative computational approach. *Microbiol Biotechnol Lett* 51:109–123. https://doi.org/10.48022/mbl.2210.10008

Kulshrestha S, Narad P, Pai SS, et al (2024) Metagenomic investigation of 16S rRNA marker gene samples to analyze the role of race, ethnicity, and location in preterm birth: a comprehensive vaginal microbiome meta-analysis. *Hum Gene* 39:201260. https://doi.org/10.1016/j.humgen.2024.201260

Kumari R, Kaur P, Verma SK, et al (2024) Omics-based cutting-edge technologies for identifying predictive biomarkers to measure the impact of air borne particulate matter exposure on male reproductive health. *JRHM* 5:2. https://doi.org/10.25259/JRHM_25_2023

Leisegang K, Sengupta P, Agarwal A, Henkel R (2021) Obesity and male infertility: mechanisms and management. *Andrologia* 53. https://doi.org/10.1111/and.13617

Liang Z, Di N, Li L, Yang D (2021) Gut microbiota alterations reveal potential gut–brain axis changes in polycystic ovary syndrome. *J Endocrinol Invest* 44:1727–1737. https://doi.org/10.1007/s40618-020-01481-5

Lilienthal D, Cahr M (2020) Genetic counseling and assisted reproductive technologies. *Cold Spring Harb Perspect Med* 10:a036566. https://doi.org/10.1101/cshperspect.a036566

Linscott ML, Chung WCJ (2020) Epigenomic control of gonadotrophin-releasing hormone neurone development and hypogonadotrophic hypogonadism. *J Neuroendocrinol* 32:e12860. https://doi.org/10.1111/jne.12860

Lowe R, Shirley N, Bleackley M, et al (2017) Transcriptomics technologies. *PLoS Comput Biol* 13:e1005457. https://doi.org/10.1371/journal.pcbi.1005457

Malone JH, Oliver B (2011) Microarrays, deep sequencing and the true measure of the transcriptome. *BMC Biol* 9:34. https://doi.org/10.1186/1741-7007-9-34

Mani S, Ghosh J, Coutifaris C, et al (2020) Epigenetic changes and assisted reproductive technologies. *Epigenetics* 15:12–25. https://doi.org/10.1080/15592294.2019.1646572

Manos J (2022) The human microbiome in disease and pathology. *APMIS* 130:690–705. https://doi.org/10.1111/apm.13225

Martin-Sanchez F, Lopez-Campos G, Gray K (2014) Biomedical informatics methods for personalized medicine and participatory health. In: *Methods in Biomedical Informatics*. Elsevier, pp 347–394.

Mbizvo MT (1996) Reproductive and sexual health: a research and developmental challenge. *Cent Afr J Med* 42:80–85.

Mercuri ND, Cox BJ (2022) The need for more research into reproductive health and disease. *eLife* 11:e75061. https://doi.org/10.7554/eLife.75061

National Institute of Environmental Health and Sciences (2024) *Reproductive Health*. National Institute of Environmental Health Sciences. www.niehs.nih.gov/health/topics/conditions/repro-health. Accessed 30 April 2024.

Ozsoy AZ, Karakus N, Yigit S, et al (2016) The evaluation of IL6 and ESR1 gene polymorphisms in primary dysmenorrhea. *Immunol Invest* 45:75–86. https://doi.org/10.3109/08820139.2015.1113426

Panni S, Lovering RC, Porras P, Orchard S (2020) Non-coding RNA regulatory networks. *Biochim Biophys Acta (BBA)—Gene Regul Mech* 1863:194417. https://doi.org/10.1016/j.bbagrm.2019.194417

Park B, Khanam R, Vinayachandran V, et al (2020) Epigenetic biomarkers and preterm birth. *Environ Epigenet* 6:dvaa005. https://doi.org/10.1093/eep/dvaa005

Park H-E, Jo SH, Lee RH, et al (2023) Spatial transcriptomics: technical aspects of recent developments and their applications in neuroscience and cancer research. *Adv Sci* 10:2206939. https://doi.org/10.1002/advs.202206939

Pique-Regi R, Romero R, Tarca AL, et al (2019) Single cell transcriptional signatures of the human placenta in term and preterm parturition. *eLife* 8:e52004. https://doi.org/10.7554/eLife.52004

Practice Committee of the American Society for Reproductive Medicine (2020) Definitions of infertility and recurrent pregnancy loss: a committee opinion. *Fertil Steril* 113:533–535. https://doi.org/10.1016/j.fertnstert.2019.11.025

Rao MS, Van Vleet TR, Ciurlionis R, et al (2019) Comparison of RNA-Seq and microarray gene expression platforms for the toxicogenomic evaluation of liver from short-term rat toxicity studies. *Front Genet* 9:636. https://doi.org/10.3389/fgene.2018.00636

Reig A, Seli E (2019) The association between assisted reproductive technologies and low birth weight. *Curr Opin Obstet Gynecol* 31:183–187. https://doi.org/10.1097/GCO.0000000000000535

Sánchez-Pavón E, Mendoza H, Garcia-Ferreyra J (2022) Trisomy 21 and assisted reproductive technologies: a review. *JBRA Assist Reprod* 26: 129–141. https://doi.org/10.5935/1518-0557.20210047

Seaby EG, Pengelly RJ, Ennis S (2016) Exome sequencing explained: a practical guide to its clinical application. *Brief Funct Genom* 15:374–384. https://doi.org/10.1093/bfgp/elv054

Sun T-C, Zhang Y, Yu K, et al (2021) LncRNAs induce oxidative stress and spermatogenesis by regulating endoplasmic reticulum genes and pathways. *Aging* 13:13764–13787. https://doi.org/10.18632/aging.202971

Uffelmann E, Huang QQ, Munung NS, et al (2021) Genome-wide association studies. *Nat Rev Methods Primers* 1:59. https://doi.org/10.1038/s43586-021-00056-9

Vora N, Parker J, Mieckowski P, et al (2019) RNA-sequencing of umbilical cord blood to investigate spontaneous preterm birth: a pilot study. *AJP Rep* 09:e60–e66. https://doi.org/10.1055/s-0039-1678717

Wan P, Meng L, Huang C, et al (2021) Replication study and meta-analysis of selected genetic variants and polycystic ovary syndrome susceptibility in Asian population. *J Assist Reprod Genet* 38:2781–2789. https://doi.org/10.1007/s10815-021-02291-1

Wang X, Shi W, Zhao S, et al (2023) Whole exome sequencing in unexplained recurrent miscarriage families identified novel pathogenic genetic causes of euploid miscarriage. *Hum Reprod* 38:1003–1018. https://doi.org/10.1093/humrep/dead039

Warr A, Robert C, Hume D, et al (2015) Exome sequencing: current and future perspectives. *G3 Genes|Genomes|Genetics* 5:1543–1550. https://doi.org/10.1534/g3.115.018564

Wyrwoll MJ, Wabschke R, Röpke A, et al (2022) Analysis of copy number variation in men with non-obstructive azoospermia. *Andrology* 10:1593–1604. https://doi.org/10.1111/andr.13267

Zeng S, Chen L, Liu X, et al (2023) Single-cell multi-omics analysis reveals dysfunctional Wnt signaling of spermatogonia in non-obstructive azoospermia. *Front Endocrinol* 14:1138386. https://doi.org/10.3389/fendo.2023.1138386

Zhang X, Qazi AA, Deshmukh S, et al (2023) Single-cell RNA sequencing reveals common and unique gene expression profiles in primary CD4+ T cells latently infected with HIV under different conditions. *Front Cell Infect Microbiol* 13:1286168. https://doi.org/10.3389/fcimb.2023.1286168

Zhang Y, Ho K, Keaton JM, et al (2020) A genome-wide association study of polycystic ovary syndrome identified from electronic health records. *Am J Obst Gynecol* 223:559.e1–559.e21. https://doi.org/10.1016/j.ajog.2020.04.004

CHAPTER 4

Proteomics and Metabolomics in Reproductive Health

Harishchander Anandaram, Keshav Mittal, Payal Gupta, Sudeepti Kulshrestha, Alakto Choudhury, Muskan Syed, Priyanka Narad, Deepak Modi, Dinesh Gupta, and Abhishek Sengupta

4.1 INTRODUCTION

Proteomics is a discipline of molecular biology that systematically investigates proteins within any biological system. It includes identifying, describing, and measuring all proteins in each sample. A proteome refers to the whole set of proteins produced by an organism or in a specific cell, tissue, organelle, etc. Proteomics employs mass spectrometry (MS), gel electrophoresis, and liquid chromatography to study and understand proteins' structure, function, and interactions. Proteomics harnesses revolutionary findings to help us better understand and administer human reproductive health. For instance, the details of specific proteins related to the functioning of sperm and the maturation of eggs have helped us know what leads to infertility. Identifying biomarkers associated with hormone regulation during pregnancy has made early detection and monitoring of reproductive disorders possible. Also notable are developments in mass spectroscopy, as carried out by John Fenn and Koichi Tanaka. This has allowed accurate protein profiling to identify slight variations of the proteome that could be associated with reproductive events (Agarwal et al. 2020). These discoveries have changed the way reproductive health issues are diagnosed and treated. Just like its genomic counterpart, the Human Proteome Project has been instrumental in mapping out the whole human proteome, for that matter, from which it becomes possible to understand more about how reproductive physiology operates at a molecular level.

DOI: 10.1201/9781003487548-4

Proteomics can be viewed as a simple analysis and an integrated approach to charting complex proteomic terrain of reproductive tissues, cells, and fluids. Using detailed analysis, we reveal the protein signatures' underlying stages of reproductive processes, which provide deep insights into molecular determinants behind fertility, conception, and embryonic development. A key aspect of our examination is recognizing the protein biomarkers this molecule signposts that lead us through the maze of reproductive health. Using quantitative proteomics methods like those involving isobaric tags, we shed light on diagnosing and prognosticating conditions such as infertility, preeclampsia, and endometriosis (Panner Selvam and Agarwal 2020). The standard methodology for targeted proteomics and analysis has been depicted in Figure 4.1.

Metabolomics technology has numerous possible applications in disease research. In contrast to traditional clinical chemistry, metabolomics is "holistic perspective" research that meticulously captures an organism's properties (Naradet al. 2022a; Tandon et al. 2023). Scientists can use high-throughput analysis combined with pattern recognition to investigate the features for identifying regularities of life activities at the metabolic level, establishing a link between metabolite and phenotype (Narad et al. 2022b). These techniques are sensitive, quick, efficient, non-invasive, and precise. The advancements in metabolomics have resulted in novel approaches to assisted reproductive technology (ART) bottleneck issues; however, more research is needed to refine these methods.

4.2 BIOMARKERS FOR REPRODUCTIVE HEALTH

Entered in the world of subtle aspects related to reproductive health, biomarkers appear as a lighthouse that guides patients toward diagnostics and monitoring. This exploration investigates the deep meaning of biomarkers significant for determining reproductive health, investigating proteins as diagnostic signals, and utilizing clues from biomolecules to monitor an individual's state. As demonstrated by Luz Cardenas and Rosanna Chianese in their study, in the field of reproductive health, biomarkers are molecular signatures that indicate one's physiological processes, pathogenic conditions, or reactions following therapy (Candenas and Chianese 2020). They are helpful because they may help identify diseases, predict disease outcomes, and advise on administering treatment. human chorionic gonadotropin (hCG) is essential in early pregnancy detection as it forms the bedrock of most tests for pregnancies. After conception, its levels go up fast, making it an ideal biomarker for pregnancy confirmation. In terms of female reproductive health, anti-Müllerian hormone (AMH) serves as a marker for ovarian reserve to provide information on potential fertility in women. Decreased AMH levels can indicate low ovarian reserve and possible fertility problems. Ovaries produce Inhibin B; this hormone is pivotal in controlling the menstrual cycle and mirrors the function of ovaries. Monitoring its levels helps to determine reproductive health, especially polycystic ovarian syndrome (PCOS). Having a central position in early pregnancy detection, the hCG is a glycoprotein formed by the development in the placenta. The precise measurement of hCG levels in maternal serum or urine allows for the early confirmation of pregnancy. AMH is the

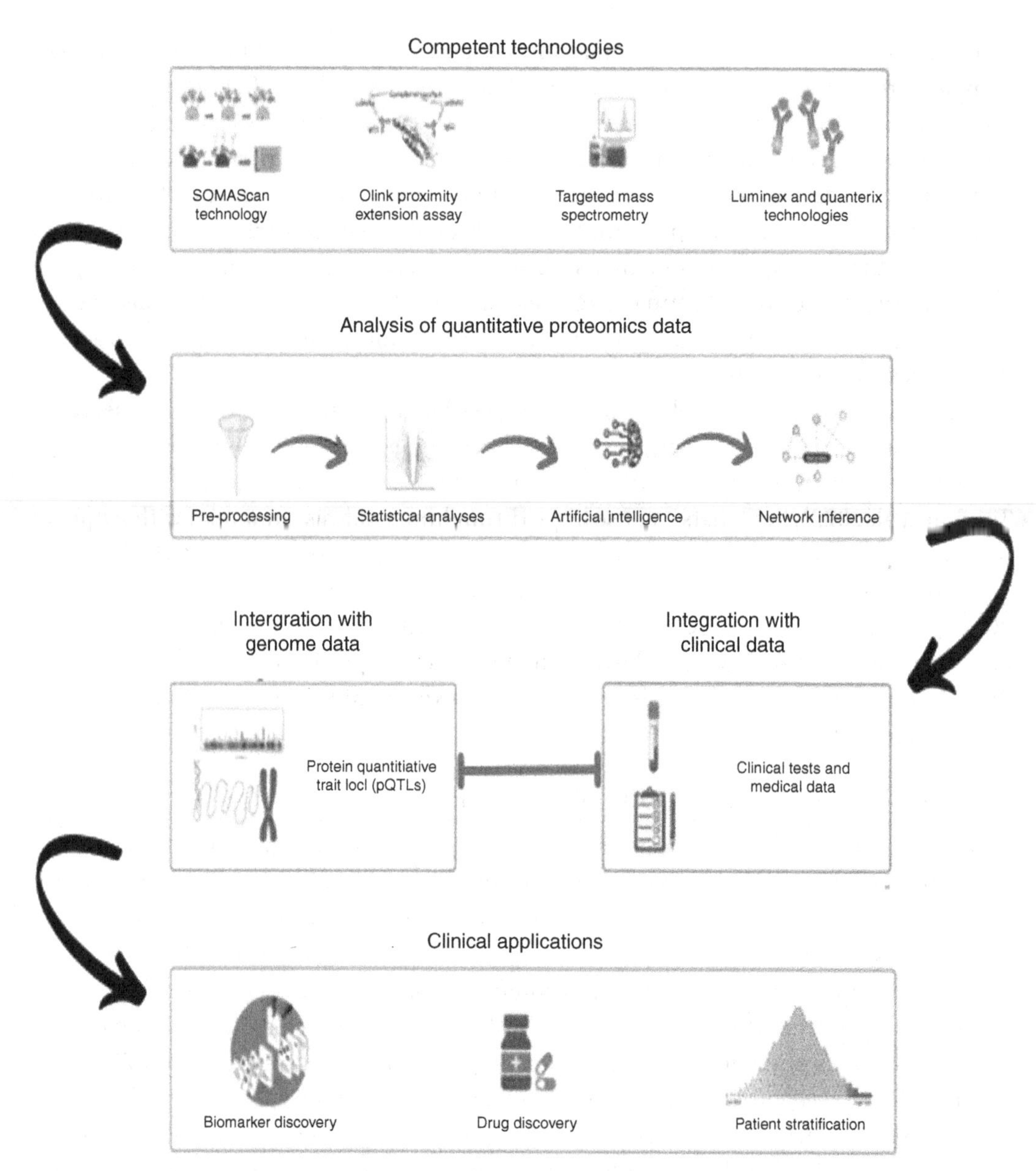

FIGURE 4.1 The standard method for proteomics. The illustration depicts various targeted technologies and the most frequent approaches for analyzing quantitative proteomics data.

main star of ovarian reserve tests, which quantitatively assesses available egg supply. Clinicians use AMH levels to facilitate conversations about fertility potential and reproductive planning. Changes in levels of Inhibin B indicate disturbances with the balance between reproductive hormones, helping to identify conditions like PCOS (Soboleva et al. 2017).

Biomarkers are crucial in the field of ART. Establishing hormonal biomarkers such as estradiol and progesterone is critical in giving valuable information about the success of fertility treatment processes, including in vitro fertilization (IVF). Proteomic profiling

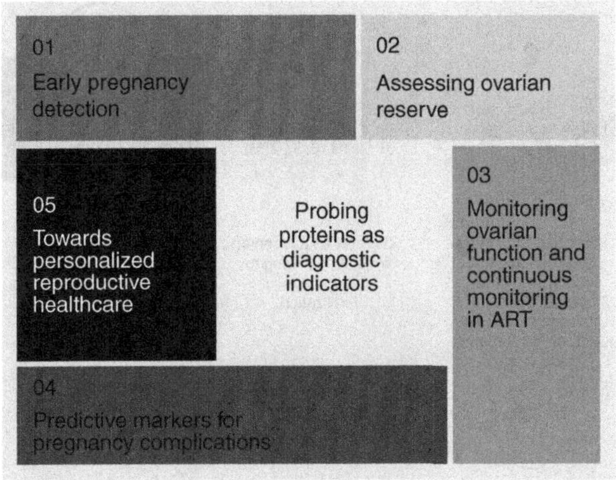

FIGURE 4.2 Illustrates biomarkers' pivotal role in reproductive health, focusing on critical diagnostic indicators and health monitoring.

helps identify potential predictive markers of pregnancy complications. By studying maternal serum proteomes, researchers can determine which protein relates to conditions such as preeclampsia and design early intervention strategies for high-risk pregnancies. The incorporation of biomarkers into reproductive healthcare provides for personalized methods. Popularizing interventions ideal for an individual's biomolecular profile improves the precision of diagnostics, maximizes fertility treatments, and reduces potential risks related to reproductive health (Snead et al. 2014). When proteins are probed as diagnostic indicators, it enables early detection and allows clinicians to guide them in personalized and precise reproductive healthcare. Biomolecular signals that lead to continuous health monitoring forecast a new age of proactive management in reproductive well-being (Figure 4.2).

4.3 PROTEOMIC APPROACHES TO PREGNANCY

With the precarious exploration of pregnancy, proteomic methods take form as a direction that reveals the complex terrain of early gestation. LC-MS is center stage, showing a panoramic view of the proteome. This method, being of high resolution and sensitivity, closely follows up the proteins in maternal serum, plasma, and uterine fluid; this paints a clear picture regarding changes occurring in protein concentration while still at the early stages of pregnancy. Applications of proteomics are also profound when the biomarkers that signal pregnancy outcomes are identified. Quantitative proteomics turns maternal serum into a gold mine, providing the basis of protein signatures that may signal early markers for potential complications. In addition to informing prognosis, this careful analysis creates opportunities for specific interventions to enhance maternal and fetal health.

The trophoblast occupies the center of early pregnancy, commanding complex molecular change necessary for implantation and placental development. With precision, trophoblastic cells are the focus of proteomic investigations that use techniques such as

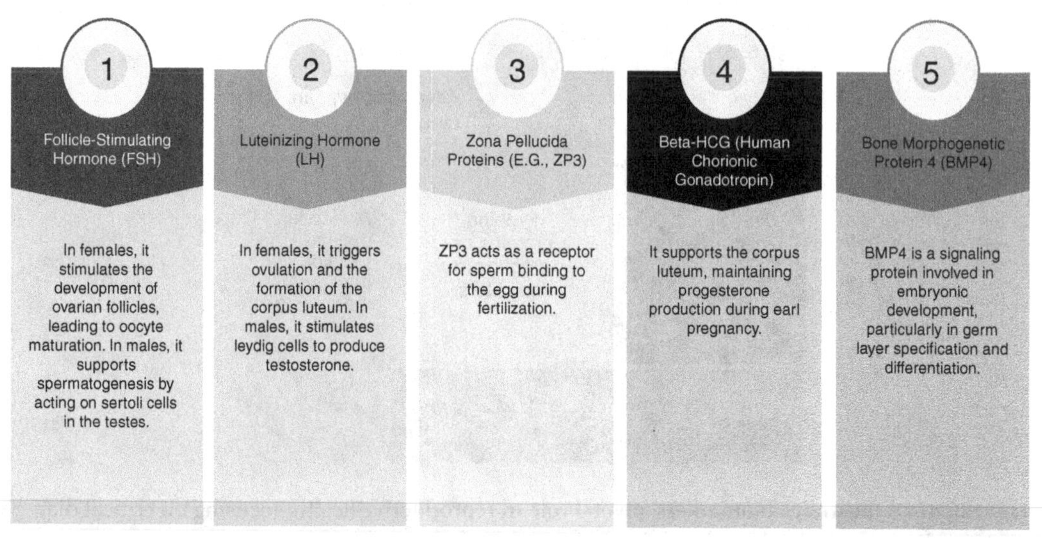

FIGURE 4.3 The five critical proteins involved in crucial reproductive processes, include gametogenesis, fertilization, and embryonic development.

two-dimensional gel electrophoresis (2D-GE) and MS. Through this granular analysis, the proteomic fingerprint of trophoblasts becomes evident. It reveals specific molecular cues directing these organisms' role in a viable establishment and sustenance during pregnancy. Advanced techniques such as isobaric tagging allow longitudinal analyses of proteomics that capture the rhythmical pattern in protein expression over pregnancy. This temporal point of view elucidates the subtle adjustments in maternal proteome that match changing needs dwelling not only on the fetus but also on actively developing itself. Maternal-placental exploration extends to the harmonious correlation between maternal and placental tissue (Miliku et al. 2015). The five critical proteins involved in crucial reproductive processes are demonstrated in Figure 4.3.

4.4 TECHNOLOGICAL ADVANCES IN REPRODUCTIVE TECHNOLOGY

Matrix-assisted laser desorption/ionization (MALDI) and electrospray ionization (ESI) techniques made the revolution in reproductive proteomics possible. These breakthroughs occurred in 1990, transforming this field by allowing efficient ionization of proteins for mass analysis. K Biemann and F Hillenkamp contributed significantly to the creation of MALDI, while JB Fellow W Vestal significantly helped with its evolution, ESI. These technologies revolutionized protein identification and quantification, creating the basis for complete proteomic analyses. Modern high-resolution imaging techniques have provided a paradigm shift in reproductive proteomics. With technologies such as MALDI imaging mass spectrometry (MADLI-IMS) and super-resolution microscopy, spatial mapping protein in reproductive tissues became possible. R M Caprioli and R L Drake first developed the system of proteomics profiling through tissue section samples using spatially resolved ionization mass spectrometry (MALDI IMS). Singe-cell proteomics represented the moving away from analysis at a population level to insights into heterogeneity inherent in cellular populations. Such revolutionary technologies as single-cell MS and microfluidic-based platforms boosted this field. Researchers like R Zenobi and J B Shear pioneered the concept of single-cell MS that involves technologies developed for analyzing an individual cell. J C Wheeler and C Offenhauser advanced microfluidic platforms or single career analysis, providing a new frontier in unraveling diversity at the cellular stage.

Reproductive biology was turned upside down by merging genomics and proteomics into quantitative proteogenomics. MS-based proteogenomic advancements have enabled genomic and proteomic information fusion, incorporating a full molecular view. Quantitative proteogenomic pioneers include S P Gygi, B W Han, and P G Schultz. Their work devising ways to associate genomic information with protein expression has laid the groundwork for a system-level understanding of reproductive processes. Rugged MS cross-linking has become a potent tool for investigating protein interactions and revealing the structure of macromolecular complexes. The methodology is characterized by establishing covalent bonds between proteins that interact with each other, allowing structural knowledge. The pioneers who partly contributed to the development of cross-linking MS include M R Chance, L Walzthoeni, and J Aside; from these, there is another name in mind associated with this technique: C R R MacQuaker. Their novel methods have opened new doors to unravel the complex protein networks that play an essential role in reproductive mechanisms.

The technological advances mark a transformative age in reproductive proteomics, allowing researchers to explore the molecular complexities underpinning fertility, conception, and gestation in unparalleled depth. These discoveries serve as pillars for unraveling the secrets of life's beginnings as the scientific community continues to push the boundaries of creativity.

4.5 MASS SPECTROMETRY AND METABOLOMICS

MS is a powerful analytical technique (Figure 4.4). Its primary role is explaining the amount and properties of biomolecules, primarily proteins, and identifying their mass-to-charge ratios (Sengupta et al. 2021). This method facilitates the identification and quantification of numerous proteins, thus decoding the intricate mosaic of molecular interactions regulating reproductive health. The power of MS in reproductive proteomics has a transformative capacity due to its capability to explore a complex world of proteins. Initially, gel electrophoresis is commonly used to separate the proteins according to their size or charge and then expose them for MS. This method extracts proteins separated on a gel and then digests enzymatically into peptides. MS is commonly used to determine the mass-to-charge ratios for these many peptides. This coupled technique, known as gel-based proteomics, allows the correlation of the gel-separated bands with MS profiles of corresponding peptides to identify and characterize them. Gel electrophoresis is used here to simplify complex mixtures of proteins before a more finely detailed and sensitive mass spectrometric analysis, ultimately increasing efficiency and accuracy in identifying the proteins within biological samples (Steen et al. 2004). The mass spectrum obtained gives an in-depth insight into proteins present within a sample. It reveals their identities

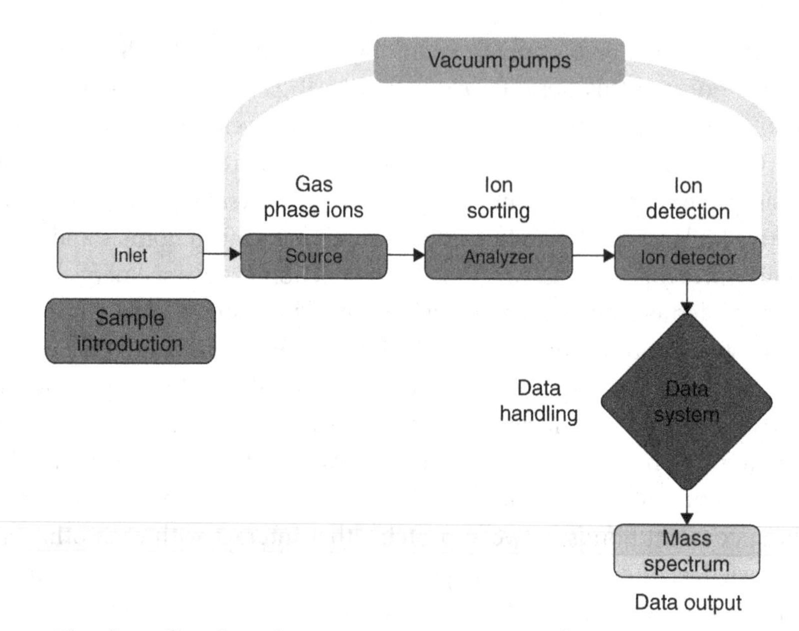

FIGURE 4.4 The above flowchart demonstrates an overview of mass spectrometry.

and how abundant they are about each other. This thoroughly explains fertility's intricacies and what happens from conception.

MS's analytical symphony in the field of reproductive health transcends mere identification. It interprets the language of proteins in seminal fluids, ovarian tissues, and maternal serum to reveal their harmonies and dissonances that lie at the basis of reproductive challenges and success. MS is the ultimate translator, turning the silent language of proteins into a brilliant story that reveals personalized interventions and improved diagnostics for reproductive medicine.

4.6 INSTRUMENTATION AND RECENT ADVANCES IN MASS SPECTROMETRY

At its core, a mass spectrometer comprises three components: (a) an ionization source, (b) a mass analyzer, and (c) a detector. It starts with ionization, where proteins are transformed into charged particles. Different ionization methods make the process versatile, including ESI and MALDI. Once ionized, the proteins pass through a mass analyzer—a magnetic or electric field where their ratios of masses to charges are scrutinized precisely. Finally, the detector records the captured ions, creating a mass spectrum revealing protein composition and concentration within an analyzed sample. Sweeping advancements in MS, as shown by Bruno Domon and Ruedi Aebersold in their work on Protein Analysis, have opened a new horizon of capabilities and possibilities that are unmatched by their previous generations to determine greater precision as well as depth proteomic analysis. One significant progress is incorporating data-independent acquisition (DIA) strategies, a novel method that improves reproducibility and reliability in protein quantification. DIA enables simultaneous fragmentation and measuring of all precursor ions, providing a broader and more consistent perspective on the proteome. Another bright spot in MS development is Parallel Reaction Monitoring (PRM) (Domon et al. 2006). This method represents a high level of sensitivity and specificity as it allows for selective analysis of specific proteins thought to be crucial.

The ability to monitor pre-defined peptide fragments makes the protein count more precise, helping decode subtle changes within the complex proteomic landscape that defines reproductive health. Innovations in high-resolution MS technologies have become instrumental and critical to progressing the limits of analytical functions. Innovations such as Orbitrap and Fourier-transform ion cyclotron resonance FT have increased mass resolution and accuracy (Sleno 2012). High-resolution MS increases the identification of proteins and allows complex protein structures to be more elucidated. Furthermore, implementing multi-omics approaches has appeared to be a revolutionary pattern. All these technological advances not only help researchers but also promise to bring transformative effects in the diagnosis, treatment, and personalized management of reproductive challenges.

4.7 DIVERSE TYPES OF MASS SPECTROMETRY

The various types of MS are shown in Figure 4.5.

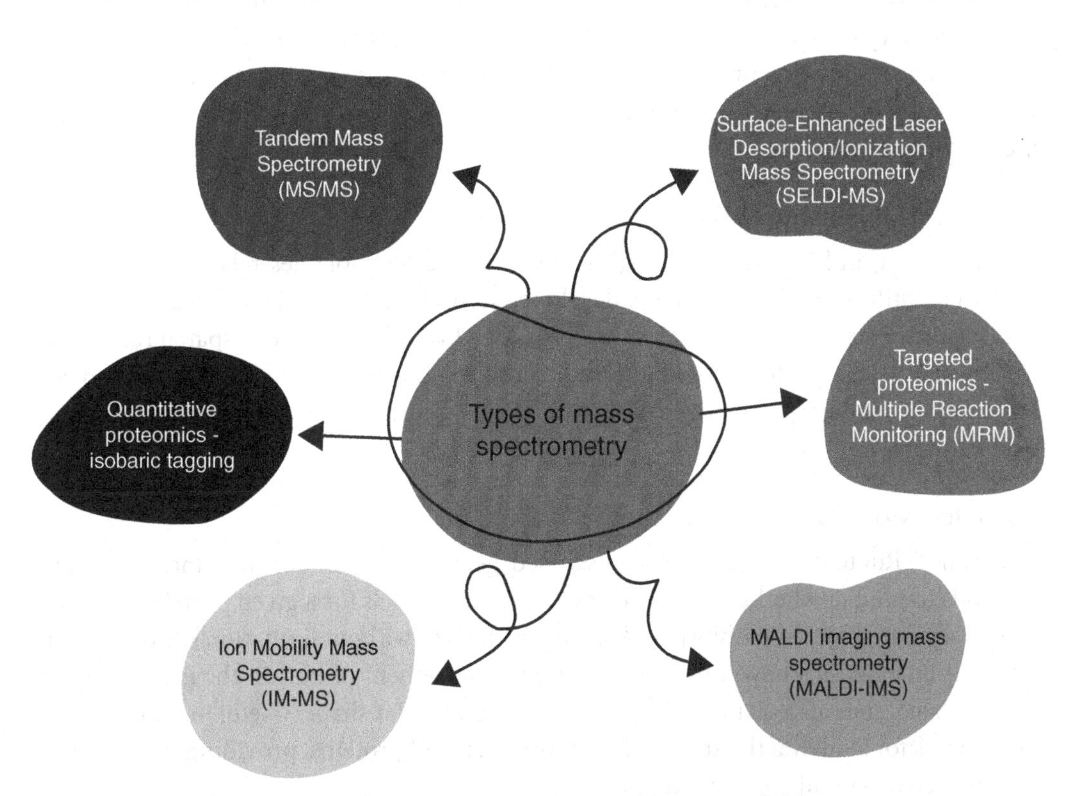

FIGURE 4.5 A diverse range of mass spectrometry is used in various fields.

4.7.1 Tandem Mass Spectrometry (MS/MS)

In his tutorial, Lekha Sleno presents the most common ion activation techniques employed in tandem MS (Sleno et al. 2004). It includes the sequential steps of ionization and mass analysis for fragmentation followed by peptide sequence. It allows detailed structural elucidation of proteins, which is essential for understanding post-translational modifications and protein interactions in reproductive proteomics.

4.7.2 Surface-Enhanced Laser Desorption/Ionization Mass Spectrometry (SELDI-MS)

In his review article, Volker Seibert addressed proteomic applications, such as target or marker identification and target validation or toxicology (Seibert et al. 2004). It uses coated surfaces to capture proteins that increase sensitivity and specificity. It helps discover biomarkers by preferentially adsorbing proteins on surfaces, allowing understanding of specific protein signatures within reproductive tissues.

4.7.3 Targeted Proteomics—Multiple Reaction Monitoring (MRM)

According to the research of Neil R Kitteringham, multiplexing and protein modification determining factors render MS an alluring alternative for antibody-based technologies; therefore, advancements in quantitative MS after the introduction of stable isotope labeling combined with scanning methods such as MRM have significantly improved both specificity and sensitivity to a level nearly equal than their immunoassay counterpart (Kitteringham et al. 2009). It targets predefined target proteins, monitoring specific precursor-to-fragment transitions. It is suitable for measuring certain proteins of interest in reproductive health studies, thus presenting accurate and focused information.

4.7.4 MALDI Imaging Mass Spectrometry (MALDI IMS)

Daniel J Ryan, in his review, highlights methods and technologies related explicitly to protein identification that have been developed to overcome these challenges in MALDI IMS experiments (Ryan et al. 2019). Combines MALDI ionization with spatial resolution MS to map tissue protein distributions. It provides spatial information about protein localization within reproductive cells, which helps understand regional protein expression patterns.

4.7.5 Ion Mobility Mass Spectrometry (IM-MS)

Brandon T Ruotolo, in their study, discussed general concepts of data interpretation, methods to predict whether some specific structural models for a given protein assembly can be separated by ion mobility, and strategies that are without limitation (Ruotolo et al. 2008). It introduces a new dimension where the movement of ions in the gas phase can be calculated, thus also giving information about molecular shape. Useful in reproductive proteomics for studying the structural conformations of proteins, providing insights into dynamic changes and conformational variables.

4.7.6 Quantitative Proteomics—Isobaric Tagging (e.g., TMT, iTRAQ)

Sivanich, Michael K, reviews several isobaric tags outlined in his review, along with recent advances such as complementary ion tags, enhancements to highly sensitive quantification of analytes that possess low abundance, and strategies designed for increasing the multiplexing capacity, which can also be observed here (Sivanich et al. 2022). The technique employs isobaric tags to label peptides, allowing the quantitation of many samples in one MS analysis. It also permits high-throughput quantitative analysis, enabling the determination of protein expression levels in various reproductive states.

4.8 TECHNOLOGICAL PLATFORMS AND TOOLS

4.8.1 Proteomics

The constant desire for precision in the field of reproductive proteomics has resulted in an abundance of creative techniques that have altered how protein analysis was previously performed. These advancements, ranging from large proteome databases to highly sophisticated tools for complicated analysis, change the scope of protein language decoding, promoting exceptional depth and precision.

Proteomic databases serve as blueprints for precision analysis. UniProt can stand as a complete protein database, offering detailed information about protein sequences, functions, and post-translational modifications. It has an extensive repository and is the primary tool for the easy identification and tagging of proteins in reproductive samples (Gomase et al. 2008). The Human Proteome Atlas maps the human proteome, comprehensively understanding protein expression patterns specific to tissues. This handy tool allows researchers to investigate the spatial arrangement of proteins in reproductive tissues and organs. PeptideAtlas is a repository of tandem MS data that helps identify and quantify peptides from various biological samples. It is critical to precision proteomic analysis of reproductive samples, improving the accuracy with which peptides can be identified is essential.

Now, we will have a look into specialized tools for proteomic analysis. The first is MaxQuant. It is a high-throughput MS data analysis computational platform that integrates sophisticated peptide identification and quantification algorithms. This tool works very well in large-scale proteomic studies of reproductive health as it offers quantitative solid data and assists protein modification identification. Another tool, Skyline, is a proteomics software that facilitates the data analysis from MRM experiments. Its roles in reproductive proteomics include enabling scientists to pinpoint specific targets and accurately measure complex biological samples. PatternLab for proteomics is an open-source program that helps analyze proteomic data through modules such as statistical validation, pattern recognition, and visualization tools (Savickas et al. 2020). This application in the field of complex proteomic datasets to enhance their interpretability by researchers helps locate patterns and trends relevant to reproductive processes.

Several advancements have been made in data processing and integration. Progenesis QI for proteomics is a software solution providing label-free quantification with detailed data processing and robust statistical analysis. Most important in reproductive proteomics, it gives a comparative study of protein expression levels to ascertain what could be biomarkers. Scaffold is a data visualization and interpretation tool that connects MS data with biological context (Correa Rojo et al. 2021). Its usage reinforces the connection between proteomic outcomes and physiological significance in reproductive studies, allowing a comprehensive understanding of complicated data sets. Collectively, these state-of-the-art tools, ranging from massive proteomic databases to specialized analysis software, drive the field of reproductive proteomics toward unparalleled accuracy. With the tools continually improving, researchers have unprecedented power to analyze and dissect the complex molecular landscape of fertility, conception, and gestation.

4.8.2 Basics of Metabolomics

There were established processes for collecting samples for metabolomics research, and the acquired metabolites were extracted using various technical procedures to examine minor statistical variations in metabolites across many groups and conduct several further investigations. There are no analytical techniques available for measuring all metabolites in human samples. Several research studies have employed diverse analyzes to perform complementary coverage. The two most popular technology platforms in metabolomics are MS and nuclear magnetic resonance (NMR).

4.8.2.1 Metabolomic Data Processing

Metabolomics technology offers numerous possible uses in illness studies. Unlike traditional clinical chemistry, metabolomics measures an organism's metabolic characteristics. Scientists utilize high-throughput analysis and pattern recognition to explore the characteristics and regularities of life activities at the metabolic level to establish a link between metabolite phenotypes. These procedures are sensitive, non-invasive, and exact. The progress of metabolomics technology has resulted in the identification of novel techniques for addressing ART bottlenecks.

4.8.2.2 Metabolomics Pretreatment

Metabolomics is used to study the metabolites of biological materials because traditional omics cannot fully explain complex physiological and pathological responses. In contrast, metabolomics aims to describe all metabolites found in cells or biological systems. Blood, urine, cerebrospinal fluid, saliva, biopsy tissue, and cell extracts are all potential materials for metabolomics study.

4.8.3 Metabolomics Applications in Reproductive Technologies

4.8..3.1 Initial Process

Metabolomics is used to study the metabolites of biological materials because traditional omics cannot fully explain complex physiological and pathological responses. In contrast, metabolomics aims to describe all metabolites found in cells or biological systems. Blood, urine, cerebrospinal fluid, saliva, biopsy tissue, and cell extracts are all potential materials for metabolomics study.

4.8.3.2 Metabolomics and Pregnancy

4.8.3.2.1 Primary Application of Metabolomics

The quality of gametes, embryos, and endometrial receptivity are three essential aspects that determine the outcome of ART, including ICSI and IVF. In clinical settings, the morphological score is used to assess the quality of gametes. When the oocyte number is low, morphology alone is insufficient for evaluation. In clinical practice, low morphological-grade oocytes can generate embryos but not live births. High-quality oocytes may fail to fertilize using ICSI, resulting in low-quality embryos that cannot implant into the endometrium. Endometrial thickness is frequently measured using ultrasonography.

4.8.3.2.2 Secondary Application of Metabolomics

Glycerylphosphorylcholine (GPC), citric acid, tyrosine, and phenylalanine could be utilized to assess sperm quality by detecting glutamine and other components in semen with NMR. This metabolomics technology offers a quick and non-invasive way to investigate infertility. Wallace et al. obtained oocytes and follicular fluids from 58 IVF patients(Wallace et al. 2012). The follicular fluids that could be fertilized but not cleaved were compared to those that could be fertilized and cleaved using one-dimensional NMR. The researchers found substantial differences in the metabolite profiles of the two forms of follicular fluid, including glucose, lactate, protein, and choline/choline phosphate levels.

4.8.4 Proteomics and Metabolomics Case Studies

Gestational diabetes mellitus (GDM) is a disorder that causes impaired glucose tolerance diagnosed during pregnancy. However, the pathophysiology of GDM is mainly unclear. GDM cases are increasing because of the worldwide obesity pandemic and maternal aging. However, GDM is anticipated to decrease following delivery. However, evidence shows that this illness is related to poor maternal and fetal outcomes, which lead to difficulties during pregnancy and birth and raise the risk of T2DM postpartum.

The work aims to develop a new differentially expressed protein (DEP) profile for GDM and normal maternal placentas and uncover putative DEPs. Research should focus on an overview of aberrant expression levels to further comprehend the complex interactions between placental proteins. However, recent GDM proteomics research focuses primarily on peripheral blood and urine samples. Two placental proteomics studies were undertaken utilizing two-dimensional electrophoresis (2-DE), matrix-assisted laser desorption/ionization time-of-flight mass spectrometry (MALDI-TOF MS), and label-free liquid chromatography with mass spectrometry elevated energy (LC-MSE). The study employed the gel-based approach to analyze and identify differential protein expression. The primary drawbacks of this approach are limited detection linearity and sensitivity, poor gel repeatability, and insufficient identification of hydrophobic or low-abundance proteins. The work by Li Ge and Pingping Huang gives more knowledge regarding the pathophysiology of GDM and is predicted to impact biomarker improvement substantially (Ge et al. 2023).

Research studies should focus on an overview of abnormal expression levels to better understand the intricate relationships between placental proteins. However, modern studies on GDM proteomics focus primarily on peripheral blood and urine samples. Two placental proteomics studies were conducted using 2-DE combined with MALDI-TOF MS and label-free LC-MSE. The study sample used the gel-based technique for differential protein expression analysis and identification. The main disadvantages of this method include low detection linearity and sensitivity, poor gel reproducibility, and insufficient identification of hydrophobic or low-abundance proteins.

4.9 CHALLENGES AND FUTURE DIRECTIONS

Overcoming the challenges in reproductive proteomics testing increasingly requires innovative solutions. This part also highlights the challenges posed in untangling human reproductive health at the molecular level, providing a detailed account of how these barriers can be overcome. In laying out the path toward future research and innovation, this analysis investigates the emerging potential that will transform our perceptions and results in reproductive health.

Proteomic approaches play a role in proteome profiling, comparative expression analysis of gene transcripts, determining posttranslational modifications, and studying protein–protein interactions. Although the human genome contains more than a thousand protein genes, it is estimated that almost a million types of proteins, including splice variants and posttranslational modifications, are produced. Proteomics provides crucial functional information on genes by the kinds of protein phosphorylation, trafficking, localization, and interactions that occur. It is not without its challenges; proteomic research is complicated by the complexity of biological samples with diverse protein concentrations, the need for adequate sample preparation procedures, advanced MS equipment, and sophisticated data processing algorithms (Agarwal et al. 2020).

The complexity of the issues that cover biological structure and physiological processes poses complications to proteomic research procedures. Observing and recognizing low-abundance proteins is difficult due to biological samples' broad range of protein concentrations. The amount of data collected with the new proteomic techniques poses problems in analyzing and processing. The quality and reproducibility of results obtained by the proteomic analysis largely depend on sample preparation. A high-end MS instrument is needed for reliable identification and characterization of proteins. More powerful software and tools are required to process and analyze high-volume proteomic data. Compared to techniques based on mass spectrometers or gel-free methods, gel methodologies have noticeable drawbacks regarding their sensitivity and reproducibility. Although reaching the qualitative differences between protein levels in biological proteomes may be difficult, recent standards such as iTRAQ provide an edge through multiplexing, quantification, and increased analytical precision and accuracy.

Developments in reproductive proteomics move toward increasing accuracy with the advantages of improved technology. Specialized diagnostic panels based on detailed proteomic profiling will likely offer better accuracy in fertility screening, gestation prenatal tests, and reproductive disorder diagnoses. The marriage between artificial intelligence and machine learning will facilitate the interpretation of protein-omics datasets, directing toward sophisticated, individualized diagnoses. Reproductive proteomics is heading to functional roles, unveiling itself from the stage of biomarker identification. The study of the dynamic protein interaction during a critical reproductive period gives much information on those mechanisms regulating reproduction health at the molecular level. Functional proteomics discovers complicated signaling pathways that expose new targets for therapeutic intervention and personalized treatments. The future is based on combining multi-omics data—genomics, transcriptomics, and metabolism. The holistic approach allows understanding of the complex and coordinated process of inter-playing molecular events, which establish reproductive outcomes. Integrating different omics datasets will enable researchers to unveil elaborate biological networks and promote holistic knowledge toward developing synergistic therapeutic approaches. Single-cell proteomic innovations deepen the understanding of cellular heterogeneity in reproductive tissues. Analyzing individual cells at the protein level reveals subtle variations hidden in mass analyses. Such an approach provides a high-resolution view of cellular diversity and differentiation abnormalities that adversely affect fertility or pregnancy complications. The development of advanced collection and analysis methods is crucial for future progress. Non-invasive and minimally invasive sampling methods and high-throughput proteomic technologies simplify data collection procedures without jeopardizing patient comfort. Advancements in microfluidic devices and innovative sample preparation approaches widen proteomic analysis efficiency in broader clinical and research settings. It is essential to address ethical issues and societal impacts. The future of reproductive proteomics must also be consciously focused on ethical frameworks that govern data privacy, consent, and findings.

4.10 CONCLUSION

In conclusion, this chapter has successfully addressed the tricky environment of human reproductive health and demonstrated a detailed account from a proteome perspective. To begin this exploration, we revealed the molecular complexity that governs reproductive structures. First, analyzing biomarkers as diagnostic signposts provided a peephole into health monitoring and an opportunity for analyzing proteins.

The rapid development of reproductive proteomics became an emphatic task, focusing on cutting-edge techniques, including MS. Such advanced technologies have enabled precision proteomic analysis, heralding a new era of understanding the reproductive health field. Employing a proteomic lens, pregnancy was scanned early in gestation, as well as maternal protein studies, for clues about sustaining a healthy pregnancy. Also, investigations into protein markers for fertility evaluation revealed molecular cues that advance our knowledge of reproductive fitness. Differentially expressed proteins in placental tissues associated with gestational diabetes provided a meaningful context for these principles, and their practical application via iTRAQ proteomics was well documented. This case shows the applied impacts of proteomic analysis to address challenges in reproductive health.

With this chapter now ending, it is essential to recognize the progress made in the field while keeping in mind some of the problems that lie within. Solving the obstacles that face reproduction proteomics requires teamwork. By planning a path forward for research and innovation, we pave the way to an advanced understanding of human reproduction. The drive toward these developments is set to define the future of this vital industry and bring hope and solutions for reproductive health.

REFERENCES

Agarwal A, Baskaran S, Panner Selvam MK, et al (2020) Unraveling the footsteps of proteomics in male reproductive research: a scientometric approach. *Antioxid Redox Signal* 32:536–549. https://doi.org/10.1089/ars.2019.7945

Candenas L, Chianese R (2020) Exosome composition and seminal plasma proteome: a promising source of biomarkers of male infertility. *IJMS* 21:7022. https://doi.org/10.3390/ijms21197022

Correa Rojo A, Heylen D, Aerts J, et al (2021) Towards building a quantitative proteomics toolbox in precision medicine: a mini-review. *Front Physiol* 12:723510. https://doi.org/10.3389/fphys.2021.723510

Domon B, Aebersold R (2006) Mass spectrometry and protein analysis. *Science* 312:212–217. https://doi.org/10.1126/science.1124619

Ge L, Huang P, Miao H, et al (2023) The new landscape of differentially expression proteins in placenta tissues of gestational diabetes based on iTRAQ proteomics. *Placenta* 131:36–48. https://doi.org/10.1016/j.placenta.2022.11.012

Gomase V, Kale K, Tagore S, Hatture S (2008) Proteomics: technologies for protein analysis. *CDM* 9:213–220. https://doi.org/10.2174/138920008783884740

Kitteringham NR, Jenkins RE, Lane CS, et al (2009) Multiple reaction monitoring for quantitative biomarker analysis in proteomics and metabolomics☆. *J Chromatogr B* 877:1229–1239. https://doi.org/10.1016/j.jchromb.2008.11.013

Miliku K, Voortman T, Van Den Hooven EH, et al (2015) First-trimester maternal protein intake and childhood kidney outcomes: the Generation R Study. *Am J Clin Nutri* 102:123–129. https://doi.org/10.3945/ajcn.114.102228

Narad P, Gupta R, Sengupta A (2022) Plant metabolomics: a new era in the advancement of agricultural research. In: *Bioinformatics in Agriculture*. Elsevier, pp 139–160.

Narad P, Naresh G, Sengupta A (2022) Metabolomics and flux balance analysis. In: *Bioinformatics*. Elsevier, pp 337–365.

Panner Selvam MK, Agarwal A (2020) Sperm and seminal plasma proteomics: molecular changes associated with varicocele-mediated male infertility. *World J Mens Health* 38:472. https://doi.org/10.5534/wjmh.190018

Ruotolo BT, Benesch JLP, Sandercock AM, et al (2008) Ion mobility–mass spectrometry analysis of large protein complexes. *Nat Protoc* 3:1139–1152. https://doi.org/10.1038/nprot.2008.78

Ryan DJ, Spraggins JM, Caprioli RM (2019) Protein identification strategies in MALDI imaging mass spectrometry: a brief review. *Curr Opin Chem Biol* 48:64–72. https://doi.org/10.1016/j.cbpa.2018.10.023

Savickas S, Kastl P, Auf Dem Keller U (2020) Combinatorial degradomics: precision tools to unveil proteolytic processes in biological systems. *Biochim Biophys Acta (BBA)—Prot Proteom* 1868:140392. https://doi.org/10.1016/j.bbapap.2020.140392

Seibert V, Wiesner A, Buschmann T, Meuer J (2004) Surface-enhanced laser desorption ionization time-of-flight mass spectrometry (SELDI TOF-MS) and ProteinChip® technology in proteomics research. *Pathol—Res Pract* 200:83–94. https://doi.org/10.1016/j.prp.2004.01.010

Sengupta A, Naresh G, Mishra A, et al (2021) Proteome analysis using machine learning approaches and its applications to diseases. In: *Advances in Protein Chemistry and Structural Biology*. Elsevier, pp 161–216.

Sivanich MK, Gu T, Tabang DN, Li L (2022) Recent advances in isobaric labeling and applications in quantitative proteomics. *Proteomics* 22:2100256. https://doi.org/10.1002/pmic.202100256

Sleno L (2012) The use of mass defect in modern mass spectrometry. *J Mass Spectrom* 47:226–236. https://doi.org/10.1002/jms.2953

Sleno L, Volmer DA (2004) Ion activation methods for tandem mass spectrometry. *J Mass Spectrom* 39:1091–1112. https://doi.org/10.1002/jms.703

Snead MC, Black CM, Kourtis AP (2014) The use of biomarkers of semen exposure in sexual and reproductive health studies. *J Women's Health* 23:787–791. https://doi.org/10.1089/jwh.2014.5018

Soboleva A, Vikhnina M, Grishina T, Frolov A (2017) Probing protein glycation by chromatography and mass spectrometry: analysis of glycation adducts. *IJMS* 18:2557. https://doi.org/10.3390/ijms18122557

Steen H, Mann M (2004) The abc's (and xyz's) of peptide sequencing. *Nat Rev Mol Cell Biol* 5:699–711. https://doi.org/10.1038/nrm1468

Tandon A, Gupta T, Verma A, et al (2023) Precision at its core: machine learning-infused metabolomics model for preterm birth prediction in human. In: *2023 14th International Conference on Computing Communication and Networking Technologies (ICCCNT)*. IEEE, Delhi, India, pp 1–7.

Wallace M, Cottell E, Gibney MJ, et al (2012) An investigation into the relationship between the metabolic profile of follicular fluid, oocyte developmental potential, and implantation outcome. *Fertil Steril* 97:10781084.e8. https://doi.org/10.1016/j.fertnstert.2012.01.122

CHAPTER 5

Systems Biology Approaches to Reproductive Health

Daniel Alex Anand and Swetha Sunkar

5.1 INTRODUCTION TO SYSTEMS BIOLOGY IN REPRODUCTIVE HEALTH

Systems biology stands as an integrative discipline that combines computational and experimental approaches to understand complex biological systems. In the context of reproductive health, systems biology offers a holistic framework to explore the multifaceted biological processes underpinning fertility, pregnancy, and reproductive disorders. This introduction aims to elucidate how systems biology principles and methodologies are being applied in the study of reproductive health, highlighting key concepts, technologies, and their transformative potential.

At its core, systems biology seeks to unravel the complexity of biological systems through an integrative approach, considering the interactions between various components of a system—genes, proteins, cells, and tissues—and how these interactions give rise to the function and behavior of that system. Unlike traditional biological research, which often focuses on isolated parts of a system, systems biology emphasizes the interconnections and dynamics within the system, aiming for a comprehensive understanding of biological phenomena (Aderem 2005).

In reproductive health, systems biology approaches have been pivotal in advancing our understanding of the reproductive system's complexity. This includes studying the hormonal regulation of the menstrual cycle, the intricacies of gametogenesis, fertilization, embryo development, and the complex interplay between genetic, environmental, and lifestyle factors affecting fertility and pregnancy outcomes.

DOI: 10.1201/9781003487548-5

Systems biology employs a variety of technologies and methodologies to study reproductive health.

- **Omics Technologies:** Genomics, transcriptomics, proteomics, and metabolomics provide comprehensive data on the genes, messenger RNA (mRNA), proteins, and metabolites involved in reproductive processes. These technologies have enabled the identification of biomarkers for reproductive disorders and insights into the molecular mechanisms underlying fertility and pregnancy (Shendure and Ji 2008).
- **Bioinformatics and Computational Modeling:** Bioinformatics tools and computational models are used to analyze and integrate omics data, facilitating the identification of key regulatory networks and pathways. Computational modeling of reproductive processes allows for the simulation of complex biological phenomena, offering insights into system behavior under various conditions.
- **Systems Pharmacology:** This approach is used to understand the effects of drugs on reproductive health at a systems level, aiding in the development of targeted therapies for reproductive disorders (Hopkins 2008).

5.1.1 Impact on Reproductive Health Research

The application of systems biology in reproductive health research has led to several significant advancements, as depicted in Table 5.1:

- **Enhanced Understanding of Reproductive Disorders:** Systems biology approaches have provided deeper insights into the etiology and progression of reproductive disorders such as polycystic ovary syndrome (PCOS), endometriosis, and infertility. By unraveling the complex molecular and cellular networks involved, researchers can identify potential targets for therapeutic intervention.
- **Personalized Medicine:** The comprehensive data generated by omics studies enable the development of personalized medicine approaches in reproductive health, allowing for tailored treatment strategies based on an individual's genetic makeup and the specific characteristics of their disorder (Collins et al. 2015).
- **Predictive Modeling:** Computational models that simulate reproductive processes can predict outcomes of fertility treatments, the impact of environmental and lifestyle factors on reproductive health, and the risk of reproductive disorders. These models are invaluable tools for clinicians and researchers, guiding decision-making and preventive strategies (Fu et al. 2022).

Despite its promise, the application of systems biology in reproductive health faces several challenges, including the complexity of reproductive systems, the need for high-quality, integrative datasets, and the development of sophisticated models that accurately reflect biological reality. Future directions will likely focus on overcoming these challenges

TABLE 5.1 Application of systems biology in reproductive health, the methods used, and their clinical impacts.

Area of Focus	Description	Key Technologies and Methodologies	Impacts and Advancements
Principles	Systems biology integrates computational and experimental approaches to understand complex biological systems holistically, emphasizing interconnections within these systems.	Integrative analysis, computational simulations	Provides a comprehensive understanding of biological phenomena
Reproductive Health Applications	It provides insights into hormonal regulation, gametogenesis, fertilization, embryo development, and the effects of genetic, environmental, and lifestyle factors on fertility and pregnancy.	Systems analysis, hormonal and genetic mapping	Advances our understanding of the reproductive system's complexity
Omics Technologies	Use of genomics, transcriptomics, proteomics, and metabolomics to gain comprehensive data on reproductive processes.	Genomic sequencing, protein profiling, metabolic mapping	Identification of biomarkers for reproductive disorders; molecular insights into fertility and pregnancy
Bioinformatics and Computational Modeling	Analyzing and integrating omics data to identify regulatory networks and simulate reproductive processes.	Data mining, network analysis, process simulation	Development of computational models to predict reproductive outcomes; identification of key regulatory networks
Systems Pharmacology	Understanding drug effects on reproductive health at a systems level.	Pharmacodynamics, pharmacogenomics	Development of targeted therapies for reproductive disorders
Personalized Medicine	Utilizing omics data for tailored treatment strategies based on individual genetic profiles.	Genetic profiling, personalized treatment plans	Creation of personalized medicine approaches in reproductive health
Predictive Modeling	Simulation of reproductive processes to predict outcomes and assess the impact of various factors.	Predictive analytics, risk assessment models	Prediction of fertility treatment outcomes; assessment of environmental and lifestyle impacts on reproductive health

TABLE 5.1 (Continued)

Area of Focus	Description	Key Technologies and Methodologies	Impacts and Advancements
Challenges	The complexity of systems, need for high-quality data, and sophisticated models that reflect biological reality.	—	Development of advanced computational tools, standardization, and sharing of omics data
Future Directions	Improvement of computational tools, data sharing, and collaborations to bridge gaps between systems biology and clinical practice.	Collaborative platforms, advanced analytics	Advancement in diagnostics, targeted therapies, and preventive strategies in reproductive health
Paradigm Shift	Represents a shift towards an integrative understanding and treatment of reproductive health issues.	Holistic research methodologies	Potential to significantly advance reproductive health research and treatment strategies

through the development of more advanced computational tools, the standardization and sharing of omics data, and interdisciplinary collaborations that bridge the gap between systems biology and clinical practice (Hood and Flores 2012).

Systems biology represents a paradigm shift in our approach to understanding and treating reproductive health issues. By leveraging the power of integrative, computational, and experimental methodologies, researchers can gain unprecedented insights into the complex biological systems at play. As technologies advance and our understanding deepens, systems biology holds the promise of significantly advancing reproductive health research, offering hope for improved diagnostic, therapeutic, and preventive strategies. This holistic approach not only enriches our scientific knowledge but also paves the way for personalized and precision medicine in reproductive health care, ultimately enhancing the health and well-being of individuals worldwide.

5.2 GENOMIC AND PROTEOMIC ANALYSES IN REPRODUCTIVE DISORDERS

To realize the potential of systems biology in the context of reproductive health, a more focused examination of the underlying molecular components is critical. Genomic and proteomic analyses serve as foundational tools in this endeavor, particularly in unraveling the complexities of disorders like infertility, PCOS), and endometriosis. These high-throughput technologies provide unparalleled insights into the genetic and protein landscapes that characterize these conditions, allowing for a deeper understanding of their etiology. They enable researchers and clinicians not only to improve diagnostics but also to tailor treatments more precisely to each individual's unique biological makeup.

Genomics, the study of an organism's complete set of DNA, including all of its genes, has revolutionized our approach to understanding reproductive disorders. By mapping the genomes of individuals with conditions like infertility, PCOS, and endometriosis, scientists have identified specific genetic variants and mutations that contribute to the risk and pathology of these diseases. For instance, genome-wide association studies (GWAS) have uncovered numerous loci associated with PCOS, suggesting it arises from a complex interplay of genetic factors (Day et al. 2015). Similarly, variations in genes related to hormone regulation, follicle development, and immune system function have been linked to infertility and endometriosis. These discoveries not only illuminate the biological mechanisms underlying these conditions but also guide the development of targeted therapies (Sapkota et al. 2017).

Complementing genomic studies, proteomics—the large-scale study of proteomes, including the set of expressed proteins in a cell, tissue, or organism—provides insights into the functional outcomes of genetic variations. Proteomic analyses in reproductive disorders focus on identifying and quantifying proteins involved in disease processes, revealing how alterations in protein expression, modification, and interaction contribute to disease phenotypes (Dumesic et al. 2015). In the context of reproductive health, proteomic technologies have been instrumental in discovering biomarkers for disease detection and progression. For example, altered levels of specific proteins in the blood or reproductive tissues have been identified in women with PCOS or endometriosis, offering potential diagnostic tools or therapeutic targets. Moreover, proteomic studies have shed light on the signaling pathways disrupted in these conditions, offering clues for intervention (Itze-Mayrhofer and Brem 2020). The applications of genomics and proteomics in reproductive health have been discussed in Table 5.2.

Research has shown that insulin resistance, a hallmark of PCOS, can be influenced by genetic variations that affect signaling pathways related to glucose metabolism. Studies have identified specific genetic variants associated with insulin resistance in PCOS, suggesting a complex interaction between genetics and metabolic processes in the condition. A detailed study on the genetic basis of insulin resistance in PCOS can be found in an article from *The Journal of Clinical Endocrinology & Metabolism*. It delves into how genetic variants are linked to alterations in protein signaling pathways, which play a crucial role in regulating glucose metabolism in PCOS patients. This study provides valuable insights into the complex interplay between genetics and metabolic functions in the development of PCOS-related insulin resistance (Stepto et al. 2019).

Indeed, the integration of complex multi-omics data, such as genomic and proteomic datasets, remains a challenge in the field of reproductive disorders and beyond. The heterogeneity of these disorders, along with varying individual genetic and environmental factors, adds layers of complexity to the interpretation of data. Advanced computational tools and machine learning algorithms offer promising solutions to analyze and interpret these large datasets to discover new biomarkers, understand disease mechanisms, and customize treatment strategies. Longitudinal studies are particularly valuable as they can track genomic and proteomic changes over time, shedding light on disease progression and treatment response.

TABLE 5.2 The use of genomics and proteomics in reproductive health, their outcomes, and the prospects for multi-omics integration.

Focus Area	Key Points	Technologies Involved	Findings & Applications
Genomic Analyses	Genomics has been crucial for identifying genetic variants contributing to reproductive disorders such as infertility, PCOS, and endometriosis.	GWAS, DNA sequencing	Discovery of specific genetic variants linked to disease risk and pathology; direction for targeted therapy development.
Proteomic Analyses	Proteomics complements genomics by revealing functional outcomes of genetic variations and the role of proteins in disease processes.	Protein expression profiling, Biomarker identification	Identification of disease biomarkers; insight into disrupted signaling pathways for potential intervention.
Insulin Resistance in PCOS	Genetic variations impact signaling pathways related to glucose metabolism, affecting insulin resistance in PCOS patients.	Genetic studies, Signaling pathway analysis	Link between genetic variants and metabolic processes in PCOS; identification of targets for therapeutic intervention.
Challenges in Multi-Omics	Heterogeneity of disorders and individual factors complicate data interpretation; longitudinal studies are essential for tracking disease progression.	Multi-omics integration, Longitudinal studies	Need for advanced computational tools to analyze complex datasets and understand disease mechanisms.
Multi-Omics Data Integration	Integration of multi-omics data is challenging due to the biological complexity and limitations of sequencing technologies.	Computational tools, Machine learning algorithms	Development of methods to analyze dynamic molecular expressions and the influence of the microbiome.
Machine Learning in Genomics	Machine learning helps in developing integrative methods for single-cell multi-omics, aiming to map cellular processes and disease states comprehensively.	Machine learning applications, Integrative single-cell analysis	Efforts to create comprehensive cellular process maps and understand disease states for personalized medicine.

As highlighted in recent literature, considerations for multi-omics data integration in the study of biological systems and diseases must take into account the biological complexity of organisms, the dynamic range of molecules, and the limitations of sequencing technologies. This includes recognizing the transient nature of certain molecular expressions, like mRNA, compared to the more stable expression of proteins. The influence of the microbiome on gene and protein expression is another crucial factor that can affect the host's epigenetic landscape, including DNA methylation and histone modifications, with potential implications for personalized medicine and therapeutic approaches (Argelaguet et al. 2021).

To keep pace with the advancements and to address these challenges, ongoing research efforts are leveraging machine learning applications in genetics and genomics. These efforts involve developing integrative methods for single-cell multi-omics, overcoming practical challenges in data analysis, and aiming to create comprehensive maps of cellular processes and disease states (Argelaguet et al. 2021). This analytical power, enriched by the precision of machine learning, sets the stage for the subsequent step in systems biology: the construction and utilization of computational models.

5.3 THE ESSENCE OF COMPUTATIONAL MODELING IN REPRODUCTIVE HEALTH

The intersection of computational modeling with reproductive health research represents a paradigm shift toward a more nuanced and predictive understanding of reproductive processes and disorders. By leveraging mathematical and computational models, scientists can simulate complex biological systems, offering insights into the mechanisms driving reproductive functions and the pathophysiology of disorders such as infertility, PCOS, and endometriosis. This exploration elucidates the role of computational modeling in advancing reproductive health research, directly informed by the intricate cellular maps produced through machine learning analyses.

Computational modeling is indeed a vital part of systems biology, providing a way to integrate various biological data into a coherent framework for the simulation and prediction of biological system behaviors. In the realm of reproductive health, these models are instrumental for simulating complex interactions, such as hormonal regulation, cellular interactions, and genetic factors, which all play a role in reproductive functions. By converting these complex biological systems into mathematical and computational representations, researchers can simulate scenarios to predict outcomes under various conditions, which can be crucial for understanding and potentially addressing reproductive disorders. The field uses a variety of computational tools and approaches, such as ordinary and partial differential equations, qualitative differential equations, and stochastic equations, among others, to model dynamic biological systems. These models allow for the exploration of biological pathways, their dynamics, and their response to different perturbations, like a drug treatment or gene knockout, without the limitations associated with direct experimentation due to ethical or practical reasons. To create these models, researchers often use established tools like BioModels, which provide access to published computational models of biological systems, and Systems Biology Markup Language (SBML), which is a standardized format for sharing models. The use of Petri nets and algorithms like the signaling Petri Net (SPN) further supports the dynamic modeling of pathways. However, it is important to recognize that the creation of these models requires experimentally derived rate constants, and the computational power needed to solve equations often limits the scope to smaller, well-characterized systems. Despite these challenges, the integration of these models into accessible software platforms allows for a more widespread use among biologists who may not be as familiar with complex mathematical techniques Kitano et al. (2002) and O'Hara et al. (2016).

5.3.1 Modeling Reproductive Endocrinology and Physiology

One of the primary applications of computational modeling in reproductive health is in understanding the hormonal regulation of the menstrual cycle, a complex interplay of the hypothalamic–pituitary–gonadal (HPG) axis (Clement 2016). Mathematical models have been developed to simulate the dynamics of hormone release and action, including the roles of gonadotropin-releasing hormone (GnRH), follicle-stimulating hormone (FSH), luteinizing hormone (LH), estrogen, and progesterone. These models help elucidate the timing and coordination necessary for ovulation and menstrual regulation, offering insights into disorders like PCOS, where these hormonal balances are disrupted (Wang et al. 2023).

5.3.2 Computational Models of Fertility and Infertility

Fertility and infertility represent areas where computational modeling has profound implications. Models simulating sperm motility, oocyte maturation, fertilization, and embryo development enable researchers to investigate the factors contributing to successful conception and the early stages of pregnancy. Computational modeling plays a significant role in understanding fertility and infertility, encompassing areas such as sperm motility, oocyte maturation, fertilization, and embryo development (Curchoe et al. 2020). Computational tools in reproductive medicine now include genetic testing to diagnose and manage infertility. For men, genetic causes can range from whole chromosomal aberrations like Klinefelter syndrome to partial aberrations such as microdeletions on the Y chromosome. For women, chromosomal disorders like reciprocal or Robertsonian translocations can affect fertility. Genetic tests can help personalize treatment strategies in assisted reproductive technologies (ART) such as in vitro fertilization (IVF), assessing risks of passing on genetic conditions, and choosing the best course of action for conception (Cariati et al. 2019. The importance of computational modeling in reproductive health has been discussed in Table 5.3.

5.3.3 Addressing Reproductive Disorders through Computational Models

Reproductive disorders such as endometriosis, PCOS, and infertility pose significant challenges due to their multifactorial nature. Computational models offer a unique advantage in deciphering these complexities by integrating genetic, proteomic, and environmental data to simulate disease pathophysiology (Giudice et al. 2023) For example, models of PCOS can incorporate insulin resistance, hormonal imbalances, and ovarian morphology to predict the disorder's progression and response to treatments. Similarly, endometriosis models can simulate the ectopic growth of endometrial tissue and its effects on fertility and pelvic pain, guiding therapeutic strategies (Dumesic et al. 2015).

5.3.4 Predictive Modeling for Personalized Medicine

Computational modeling in reproductive health is indeed aimed at enhancing personalized medicine. By incorporating specific patient data into models, the prediction of treatment outcomes can be refined, and the most effective interventions for the

TABLE 5.3 The role of computational modeling in reproductive health, its techniques, and its impact on research and practice.

Area	Description	Techniques	Outcomes
Computational Modeling	A vital part of systems biology, it allows for simulation and prediction of biological system behaviors in reproductive health.	Mathematical and computational models using tools like BioModels, SBML, and Petri nets.	Insights into reproductive function mechanisms and disorder pathophysiology.
Reproductive Endocrinology and Physiology	Understanding hormonal regulation of the menstrual cycle via simulation of hormone dynamics.	Mathematical models of hormone release and action (GnRH, FSH, LH, estrogen, progesterone).	Elucidation of timing and coordination for ovulation, insights into disorders like PCOS.
Fertility and Infertility	Computational models simulate aspects such as sperm motility and embryo development, assisting in understanding factors contributing to conception.	Models for gamete interaction, genetic testing in ART.	Support for treatment strategies in ART, assessment of genetic risks, and improved conception approaches.
Addressing Reproductive Disorders	Computational models integrate genetic, proteomic, and environmental data to simulate disease pathophysiology.	Disease progression and treatment response models for PCOS and endometriosis.	Predictive modeling for progression and response to treatments, guiding therapeutic strategies.
Predictive Modeling for Personalized Medicine	Aimed at enhancing personalized medicine by refining treatment outcome predictions based on specific patient data.	Data science, machine learning techniques, deep learning for large dataset analysis.	Improvement in prediction of treatment outcomes, reduced risk of adverse effects, tailored interventions.
Personalized ART Models	Personalized modeling of hormonal cycles to improve treatment success rates in ART.	Machine learning (Gaussian process regression) for hormonal value estimation.	Customized treatment protocols for ART, potentially improving treatment outcomes.
Research Directions	Improving model accuracy with machine learning algorithms and deep learning, integrating multi-omics data for a comprehensive understanding.	Deep learning for multi-omics data fusion, user-friendly clinical decision support tools.	Advances in diagnostic and treatment strategies, personalized to the patient's omics profile.

individual can be identified, reducing the risk of adverse effects. Data science, especially machine learning, is central to this approach, using techniques like deep learning to process large datasets in areas such as radiology and personalized medicine. However, challenges remain, including ensuring sufficient prediction performance for clinical practice, difficulties in interpretation, and validation for clinical use. Despite the hype, very few machine learning-based solutions currently have a clinical impact due to these challenges (Fröhlich et al. 2018).

In the realm of ART, personalized models based on individual hormonal profiles are indeed promising for improving treatment success rates. A study led by Iñigo Urteaga and colleagues at Columbia University, along with Clue in Berlin, focused on personalized modeling of the female hormonal cycle using machine learning. They investigated the use of Gaussian process regression to estimate hormonal values over time, which could predict menstrual cycle phases. This approach allows for simulations tailored to the individual's unique hormonal fluctuations, potentially improving ART outcomes by customizing treatment protocols. The study demonstrates that such personalized models could significantly advance reproductive healthcare by offering more targeted and efficient treatments (Urteaga et al. 2017)

Research in reproductive health is indeed geared toward improving model accuracy with the help of machine learning algorithms and deep learning. These technological advancements focus on integrating multi-omics data to gain a more comprehensive understanding of reproductive disorders. The development of user-friendly tools for clinical decision support is also a key part of this endeavor. Efforts to integrate multi-omics data using machine learning strategies are critical, especially given the complex interplay of genetic, environmental, and clinical factors in reproductive health outcomes.

The research in this area highlights that using deep learning for multi-omics data fusion can significantly advance our understanding of complex diseases such as cancer, which can translate to reproductive disorders as well. Such methods can be used for robust classification of disease subtypes and prediction of clinical outcomes. Moreover, the practical application of these advanced analytical tools in clinical settings has the potential to lead to improved diagnostic and treatment strategies, personalized to the individual patient's omics profile (Pammi et al. 2023, Leng et al. 2022). The transition from theoretical models to real-world clinical applications underscores the impact of machine learning on reproductive health.

5.4 THE FOUNDATION OF FERTILITY: KEY MOLECULAR PATHWAYS

5.4.1 Gametogenesis and Genetic Integrity

Delving deeper into the mechanisms of gametogenesis reveals that the precision of this process is paramount. The intricate interplay of the meiotic recombination machinery, DNA repair enzymes, and the spindle assembly checkpoint not only facilitates the accurate distribution of chromosomes but also guards against genetic anomalies. Disruptions in these crucial pathways can lead to significant reproductive challenges, such as aneuploidy and other genetic abnormalities, which are major contributors to infertility (Nagaoka et al. 2012). Understanding these complex molecular interactions is therefore not only fundamental for diagnosing reproductive health issues but also for developing targeted interventions that can mitigate these fertility barriers.

5.4.2 Hormonal Regulation and Signal Transduction in Fertility

"Molecular mechanisms of action of FSH" by Livio Casarini and Pascale Crépieux provides an in-depth review of how follicle-stimulating hormone (FSH) acts through its receptor (FSHR) to regulate reproductive functions. This hormone plays a crucial role in gametogenesis by interacting with intracellular signaling networks. The FSH-induced signaling pathways involve various molecular interactions that mediate cell proliferation and steroidogenesis in gonadal cells. These pathways are essential for reproductive development and functions such as folliculogenesis, oocyte selection, and the synthesis of sex steroid hormones, crucial for fertilization, implantation, and pregnancy (Casarini and Crépieux 2019).

5.4.3 Molecular Pathways in Infertility Pathogenesis

Infertility often arises from dysfunctions within the molecular pathways governing reproductive processes. In conditions like PCOS and endometriosis, altered hormonal signaling, inflammation, and immune responses can interfere with ovulation, fertilization, and implantation (Dumesic et al. 2015). Advances in genomic and proteomic analyses have identified specific molecular markers and pathways implicated in these disorders, offering new avenues for diagnosis and treatment.

The role of systems biology in understanding fertility pathways is depicted in Table 5.4.

5.4.4 Oxidative Stress and Reproductive Health

Oxidative stress, resulting from an imbalance between reactive oxygen species (ROS) production and antioxidant defenses, has been linked to infertility (Agarwal et al. 2019). Excessive ROS can damage DNA, proteins, and lipids in reproductive cells, impairing their function. Understanding the molecular mechanisms of oxidative stress and its impact on fertility is crucial for developing antioxidant-based therapeutic strategies.

5.4.5 Emerging Research and Therapeutic Targets

Recent research has focused on identifying novel molecular targets for treating infertility. This includes investigating the role of microRNAs in regulating gene expression related to reproductive processes, exploring the potential of stem cell therapy for restoring fertility, and targeting specific signaling pathways with pharmaceutical agents. Such approaches promise to enhance fertility treatments and improve outcomes for individuals struggling with infertility (Bahmyari et al. 2021).

While there have been significant advancements in the field of reproductive health, understanding the molecular pathways associated with fertility and infertility remains a

TABLE 5.4 The role of systems biology in understanding fertility pathways and advancing precision medicine.

Topic	Description	Molecular Pathways and Processes	Implications for Fertility
Gametogenesis and Genetic Integrity	The process ensures the production of genetically stable sperm and oocytes.	Meiotic recombination, DNA repair, spindle assembly checkpoint.	Disruptions can lead to aneuploidy and infertility.
Hormonal Regulation in Fertility	FSH's role in reproductive functions through its receptor, FSHR.	Intracellular signaling for cell proliferation and steroidogenesis in gonadal cells.	Essential for folliculogenesis, oocyte selection, and hormone synthesis.
Molecular Pathways in Infertility	Dysfunctions in pathways can lead to reproductive disorders like PCOS and endometriosis.	Altered hormonal signaling, inflammation, and immune responses.	Provides targets for new diagnostic and treatment methods.
Oxidative Stress and Fertility	The imbalance between ROS production and antioxidant defenses is linked to infertility.	DNA, protein, and lipid damage in reproductive cells from oxidative stress.	Antioxidant-based therapies could mitigate fertility issues.
Emerging Research and Therapeutic Targets	Identification of novel targets for fertility treatment.	MicroRNAs, stem cell therapy, specific signaling pathways.	Potential enhancement of fertility treatments and outcomes.
Challenges and Directions	The complexity of molecular pathways and genetic variability makes treatment development challenging.	Integration of systems biology methodologies.	Aiming toward precision medicine in reproductive health.

formidable challenge. The complexity of these pathways, combined with individual genetic variability, makes it difficult to develop universal treatments that are effective for everyone. As a result, there is a pressing need for further research, which should prioritize the integration of systems biology approaches. Such methodologies aim to map out the intricate network of interactions that govern reproductive processes, thereby paving the way for the application of precision medicine in this vital area of health.

5.5 INTRODUCTION TO EPIGENETICS IN REPRODUCTIVE HEALTH

Building on the molecular foundations, the field of epigenetics offers profound insights into reproductive health and disease. Epigenetics involves heritable changes in gene expression that occur without alterations to the underlying DNA sequence. Through mechanisms such as DNA methylation, histone modification, and the activity of non-coding RNAs, epigenetic modifications play crucial roles in controlling gene expression and cellular functions. This section explores how these changes influence fertility, affect pregnancy outcomes, and contribute to the pathogenesis of reproductive disorders. By understanding these epigenetic mechanisms, researchers and clinicians can better address the complexities of reproductive health and develop more targeted interventions.

The field of epigenetics bridges the gap between genetic predisposition and environmental influences, offering insights into how external factors can modulate gene expression to impact reproductive health. Epigenetic modifications play crucial roles in gametogenesis, embryo development, and placental function, ensuring proper gene regulation is maintained throughout these processes (Feil and Fraga 2012) Dysregulation of epigenetic mechanisms, however, can lead to altered fertility, pregnancy complications, and an increased risk of reproductive diseases (Huntriss et al. 2018).

5.5.1 Epigenetic Regulation of Gametogenesis and Embryonic Development

During gametogenesis, epigenetic reprogramming ensures the erasure and establishment of sex-specific epigenetic marks, which are critical for the development of sperm and oocytes (Morgan et al. 2005). Similarly, embryonic development requires precise epigenetic modifications to regulate gene expression patterns essential for cell differentiation and organogenesis. Errors in these processes can result in developmental abnormalities and impact the health of the offspring (Santos and Dean 2004).

5.5.2 The Impact of Epigenetics on Fertility and Infertility

Epigenetic mechanisms are intimately involved in the regulation of fertility. For instance, DNA methylation patterns in the promoters of genes controlling ovarian function and spermatogenesis can influence fertility (Stuppia et al. 2015). Alterations in these patterns have been associated with conditions such as PCOS and idiopathic male infertility, suggesting that epigenetic dysregulation may underlie some cases of reproductive dysfunction (Kobayashi et al. 2007).

5.5.3 Epigenetic Contributions to Reproductive Diseases

Research in reproductive science indicates that DNA methylation, a critical epigenetic mechanism, plays a significant role in regulating gene expression as discussed in Table 5.5. Abnormalities in DNA methylation patterns have been associated with various reproductive disorders, including endometriosis. These epigenetic changes can affect genes involved in immune responses and inflammation, which may contribute to the pathology of the disease. The study of epigenetic alterations in reproductive diseases like endometriosis is complex, with various factors influencing gene regulation through methylation changes (Calicchio et al. 2014). Understanding these epigenetic landscapes opens new avenues for diagnosis and therapeutic intervention.

5.5.4 Environmental Factors, Lifestyle, and Epigenetic Modifications

Epigenetic modifications can also be influenced by environmental factors and lifestyle choices, including diet, physical activity, and exposure to toxins (Jirtle and Skinner 2007). These modifications may not only affect the individual's reproductive health

TABLE 5.5 Epigenetic influences on fertility, infertility, oxidative stress impacts, and therapeutic innovations in reproductive health.

Area of Impact	Role of Epigenetics	Key Mechanisms	Implications for Research and Treatment
Gametogenesis and Genetic Integrity	Ensures proper chromosomal segregation and genetic stability during gamete formation	DNA methylation, histone modification	Identification of genetic abnormalities contributing to infertility
Hormonal Regulation and Fertility	Influences fertility and pregnancy outcomes	FSH signaling pathways through receptor interactions	New avenues for diagnosis and treatment of fertility issues
Infertility Pathogenesis	Altered molecular pathways lead to conditions like PCOS and endometriosis	Identification of specific molecular markers and pathways	Advanced genomic and proteomic analyses to guide therapeutic interventions
Oxidative Stress	Impacts fertility through DNA, protein, and lipid damage in reproductive cells	Balance between ROS production and antioxidant defenses	Antioxidant-based therapeutic strategies for infertility
Emerging Therapeutic Targets	MicroRNAs, stem cell therapy, and pharmaceutical agents targeting signaling pathways	Regulation of gene expression related to reproductive processes	Enhancement of fertility treatments and improved outcomes

but can also be passed on to offspring, potentially impacting their health and susceptibility to diseases (Donkin and Barres 2018). This highlights the importance of considering epigenetic factors in public health strategies for improving reproductive outcomes.

5.5.5 Emerging Therapies Targeting Epigenetic Mechanisms

The reversible nature of epigenetic modifications presents unique opportunities for therapeutic intervention. Drugs targeting specific epigenetic enzymes, such as DNA methyltransferases and histone deacetylases, are being explored for the treatment of reproductive disorders (Giacone et al. 2019). Additionally, lifestyle interventions aimed at modulating the epigenome offer a promising avenue for improving reproductive health and preventing disease transmission to future generations (Li 2018).

Significant advances have been made in the realm of reproductive health, yet full comprehension of the epigenome's complexity remains a challenge. The epigenome is a critical facet of genetic regulation with significant implications for reproductive health, as noted by Feinberg et al. (2016). To unravel these complex regulatory networks, the development of sophisticated epigenomic technologies and bioinformatics tools is vital. Future research is mandated to pinpoint the precise mechanisms by which epigenetic modifications affect reproductive processes. Understanding these mechanisms is essential to translating epigenetic insights into practical applications in clinical settings, a point emphasized by Rando and Simmons (2015).

5.6 SYSTEMS BIOLOGY IN PRENATAL DEVELOPMENT AND PREGNANCY

The field of systems biology stands out for its comprehensive approach to deciphering biological complexity, significantly enriching our understanding of prenatal development, pregnancy, and fetal health. This discipline integrates computational models with experimental data, shedding light on the intricate networks of genes, proteins, and environmental factors that influence the development and sustenance of pregnancy, as well as the onset of related complications. This section will delve into the application of systems biology methodologies to demystify the complexities of prenatal development and pregnancy. It will highlight how these methodologies can inform our understanding of fetal health, uncover the mechanisms at the root of pregnancy complications, and guide the development of potential therapeutic interventions.

5.6.1 Integrative Modeling of Prenatal Development

Prenatal development is a highly orchestrated process involving gene expression regulation, signaling pathway interactions, and cellular differentiation and growth. Systems biology approaches employ computational modeling to integrate genomic, epigenomic, and proteomic data, creating comprehensive models that simulate the molecular dynamics of embryogenesis (Hood and Galas 2003). These models can predict developmental outcomes based on genetic and environmental variables, providing invaluable insights into the molecular basis of congenital anomalies and the effects of external factors on fetal development (Elmannai et al. 2023).

5.6.2 Deciphering Pregnancy Complications

Pregnancy complications such as preeclampsia, gestational diabetes, and preterm birth pose significant risks to both mothers and fetuses. Systems biology has been instrumental in identifying the complex biological networks and pathways associated with these conditions. For instance, network analysis of gene expression and signaling pathways has revealed key molecular players in the pathophysiology of preeclampsia, suggesting novel biomarkers for early detection and potential therapeutic targets (Levine et al. 2006). Similarly, metabolic network modeling offers a deeper understanding of gestational diabetes, highlighting the role of glucose metabolism and insulin signaling alterations (Lowe and Karban 2014).

5.6.3 Fetal Health and Growth Monitoring

Systems biology and machine learning are being increasingly utilized to monitor and predict fetal health and growth by analyzing data from omics technologies alongside clinical observations, as depicted in Table 5.6. These tools can detect patterns in vast data sets to pinpoint markers for conditions such as fetal distress and growth restrictions. Recent studies have made strides in understanding the healthy maternal–fetal interface and have applied novel approaches for predicting gestational age and changes associated

TABLE 5.6 Systems biology in prenatal care, its methods, and benefits for fetal health management.

Research Area	Systems Biology Application	Key Methods	Implications
Prenatal Development	Computational modeling integrates multi-omics data to simulate embryogenesis.	Genomic, epigenomic, and proteomic integration	Predicts developmental outcomes, understanding congenital anomalies
Pregnancy Complications	Identifies biological networks and pathways involved in conditions like preeclampsia and gestational diabetes.	Network analysis, metabolic network modeling	Novel biomarkers for early detection, potential therapeutic targets
Fetal Health and Growth	Applies machine learning to omics and clinical data to monitor fetal health.	Data pattern analysis, multi-omics data integration	Early intervention strategies, improved prenatal care
Placental Function	Studies placental genomics to understand its role in health traits and diseases.	Genomic analysis	Insight into molecular underpinnings of diseases, precise medical interventions
Maternal–Fetal Interface	Develops gene expression references for healthy placental tissue.	Single-cell RNA-Seq, gene expression deconvolution	Advances in molecular epidemiology, understanding of pregnancy complications

with pregnancy through multi-omics data integration. This integrated approach is helping to map fetal and neonatal immune system development, potentially leading to improvements in predicting and managing conditions affecting fetal health (Ozen et al. 2023), or anomalies (Mennickent et al. 2023). This approach enables early intervention strategies that can mitigate adverse outcomes, enhancing prenatal care and management.

5.6.4 Placental Function and Maternal–Fetal Interface

A comprehensive reference that explores the role of placental genomics in health traits and diseases is provided in a study published in *Nature Communications*. The research presents the intricate interplay between genetic factors and placental function, highlighting how genomics can mediate associations with various complex health conditions. The study delves into the importance of placental genomics for understanding the molecular underpinnings of diseases, potentially paving the way for more precise medical interventions (Bhattacharya et al. 2022).

Another valuable study published in *Communications Biology* develops a reference for gene expression deconvolution in healthy placental villous tissue. This work is significant for advancing the field of perinatal molecular epidemiology, helping to understand the role of different cell types and their proportions in the placenta, and how these may influence conditions such as preeclampsia. The research underscores the power of using single-cell RNA-sequencing data to decode the complex cellular architecture of the placenta, which is essential for comprehending various pregnancy-related complications (Campbell et al. 2023).

5.6.5 Environmental Influences on Pregnancy Outcomes

Environmental factors, including nutrition, stress, and exposure to toxins, indeed play a significant role in pregnancy outcomes. Various studies have indicated that prenatal exposure to environmental contaminants may affect the future health of both the mother and the child. Critical periods such as pregnancy and early life are particularly sensitive to these exposures. For instance, the consumption of alcohol during pregnancy can lead to fetal alcohol spectrum disorders, with fetal alcohol syndrome being the most severe form. Smoking is linked to several negative outcomes, such as placental abnormalities, preterm birth, stillbirth, and later in life, the risk of intellectual impairment, obesity, and cardiovascular diseases. The use of drugs can also result in negative birth outcomes.

Pregnant and lactating women are exposed to endocrine-disrupting chemicals and heavy metals in food, which may alter hormone levels in the body and have been associated with conditions like preeclampsia and intrauterine growth restriction. Metals can accumulate in the placenta and lead to fetal growth restriction. The impact of air pollutants is also significant, with growing evidence linking it to outcomes like preterm birth, fetal growth restriction, gestational diabetes, and reduced telomere length in infants. While breastfeeding has many advantages, it is also a pathway for contaminants to pass to the infant, although the health benefits of breastfeeding generally outweigh the risks from these contaminants.

Additionally, initial studies have suggested an association between preeclampsia and environmental noise, particularly early-onset preeclampsia. This growing evidence of the adverse effects of environmental contaminants, particularly following exposure during pregnancy and breastfeeding, highlights the need for it to be considered a major public health issue (Gómez-Roig et al. 2021). Systems biology approaches model the interactions between these environmental factors and biological systems to understand their effects on pregnancy and fetal development. Such models can guide public health recommendations and interventions to minimize environmental risks during pregnancy (Padula et al. 2020).

Systems biology studies focusing on prenatal development and pregnancy are confronted with the need for extensive, high-quality omics data and the integration of varied data types. Moving forward, progress in this field will hinge on the creation of more sophisticated computational models, the enhancement of data collection and sharing practices, and the strengthening of interdisciplinary collaborations. There is also an imperative need to address ethical considerations when applying systems biology findings in clinical contexts, to ensure that any interventions derived from these insights are both safe and beneficial for patients.

5.7 BIOINFORMATICS AND ITS IMPACT ON REPRODUCTIVE HEALTH RESEARCH

Bioinformatics stands at the intersection of biology, computer science, mathematics, and statistics, and is essential for the analysis and interpretation of the immense data pools produced by contemporary biological and clinical studies. Within the sphere of reproductive health, bioinformatics tools are invaluable for navigating through expansive genomic, transcriptomic, and metabolomic datasets. These tools aid in unraveling complex insights related to fertility, pregnancy, and reproductive pathologies. This section will examine how bioinformatics is applied in reproductive health research, highlighting its transformative impact on our comprehension of biological intricacies, the diagnosis of various conditions, and the sculpting of new therapeutic avenues.

The impact of high-throughput sequencing (HTS) is well-documented across various fields, including reproductive health. HTS has transformed genomic research by enabling the study of genetic information at a high resolution in a cost-effective manner. Innovations by companies like Illumina, PacBio, and Oxford Nanopore have greatly improved the accuracy and length of sequencing, making vast genomic data more accessible and useful for medical research and diagnostics. The adoption of these technologies by healthcare and governmental agencies illustrates their significant role in advancing public health (Imanian et al. 2022).

5.7.1 Genomic Data Analysis in Reproductive Health

Genomic data analysis, facilitated by bioinformatics, has led to significant breakthroughs in identifying genetic predispositions to conditions such as PCOS, endometriosis, and idiopathic infertility. By employing algorithms for variant calling, GWAS, and linkage analysis, researchers can pinpoint genetic variations that contribute to the susceptibility and pathogenesis of reproductive diseases (Nautiyal et al. 2022, McCarthy et al. 2008). These insights are critical for developing targeted therapeutic interventions and personalized treatment plans (Dunaif and Fauser 2013). The role of bioinformatics in reproductive health data analysis has been discussed in Table 5.7.

5.7.2 Transcriptomic Insights into Fertility and Pregnancy

Transcriptomics, the study of the complete set of RNA transcripts produced by the genome, offers a dynamic view of gene expression changes during different stages of reproductive processes. Bioinformatics tools analyze RNA sequencing (RNA-seq) data to elucidate the transcriptional profiles associated with normal fertility, pregnancy, and complications such as preeclampsia and recurrent pregnancy loss (Buckberry et al. 2017). Understanding these transcriptional landscapes helps uncover the regulatory networks controlling reproductive functions and identifying potential targets for therapeutic intervention (Suhorutshenko et al. 2018).

5.7.3 Metabolomic Approaches to Understanding Reproductive Disorders

Metabolomics, the comprehensive analysis of small molecule metabolites within cells, tissues, or organisms, provides insights into the metabolic alterations associated with reproductive health and disease (Kenny et al. 2010). Bioinformatics plays a key role in metabolomic data analysis, enabling the identification of metabolic signatures of diseases like PCOS and gestational diabetes. These metabolic profiles offer clues to the underlying

TABLE 5.7 Bioinformatics in reproductive health data analysis, its achievements, challenges, and research prospects.

Field of Study	Bioinformatics Applications	Results	Challenges and Future Directions
Genomic Data Analysis	Utilizes algorithms for GWAS, variant calling, linkage analysis to identify genetic variations.	Breakthroughs in identifying genetic predispositions to reproductive diseases like PCOS and endometriosis.	Need for advanced algorithms, handling complex datasets, and ensuring data privacy.
Transcriptomic Insights	Analyses RNA-Seq data to study gene expression during reproductive processes.	Insights into regulatory networks for fertility, pregnancy, and complications like preeclampsia.	Integration of data to fully understand gene expression changes and impacts.
Metabolomic Approaches	Employs data analysis for identifying metabolic signatures of diseases.	Understanding of metabolic alterations in reproductive disorders for diagnostics and treatment.	Sophisticated tools for metabolomic data analysis and interpretation.
Integrative Bioinformatics	Combines genomics, transcriptomics, and metabolomics for a systems biology view.	Comprehensive models of molecular pathways affecting fertility and reproductive disorders.	Development of robust computational resources and improved databases for holistic analysis.

pathophysiology of reproductive disorders and can guide the development of novel diagnostics and treatments (Escobar-Morreale 2018).

5.7.4 Integrative Bioinformatics for Holistic Understanding

An integrative bioinformatics approach, combining data from genomics, transcriptomics, and metabolomics, offers a holistic view of the biological systems involved in reproductive health (Karczewski and Snyder 2018). By integrating and interpreting data across these omics layers, researchers can construct comprehensive models of the molecular pathways affecting fertility, pregnancy outcomes, and the development of reproductive disorders (Joyce and Palsson 2006). This systems biology perspective is crucial for identifying biomarkers, unraveling disease mechanisms, and designing effective therapies.

Bioinformatics offers immense potential in reproductive health research, yet its application is not without challenges. It requires robust computational resources and sophisticated algorithms capable of integrating multifaceted data streams. Managing large and complex datasets while ensuring data privacy and security is another significant hurdle, as identified by Horgan and Kenny (2011). Future research is set to focus on the development of advanced bioinformatics tools, the creation of improved databases specific to reproductive health, and harnessing the power of artificial intelligence and machine learning to predict disease risks and treatment outcomes with greater precision.

5.8 FOUNDATIONS OF PERSONALIZED MEDICINE IN REPRODUCTIVE HEALTH

Personalized medicine marks a revolutionary shift in reproductive health, offering customized prevention, diagnosis, and treatments based on an individual's specific biological markers, lifestyle, and environmental factors. At the forefront of this movement is systems biology, which provides a comprehensive framework for integrating data from genomics, proteomics, metabolomics, and bioinformatics. Such a holistic approach is essential for dissecting the subtle variations that influence reproductive health issues, thereby enabling precise and personalized interventions. This section will delve into how systems biology contributes to the evolution of personalized medicine within the field of reproductive health. It will detail the applications of this approach in enhancing fertility treatments, optimizing pregnancy management, and improving the treatment of various reproductive disorders.

Personalized medicine in reproductive health leverages individual genetic, epigenetic, and omics profiles to guide clinical decisions. Systems biology supports this approach by unraveling the intricate web of biological interactions underlying reproductive functions and disorders. Through comprehensive models that consider genetic predispositions and environmental influences, systems biology helps predict individual responses to treatments, enhancing the efficacy and safety of fertility interventions and pregnancy management.

5.8.1 Enhancing Fertility Treatments through Systems Biology

In fertility treatments, such as IVF, systems biology approaches analyze patient-specific data to optimize protocols and improve success rates. By examining genetic markers, hormonal levels, and other omics data, clinicians can tailor stimulation protocols and embryo selection processes to the individual's unique profile, significantly increasing the chances of conception and healthy pregnancy. Additionally, systems biology can identify biomarkers predictive of ovarian hyperstimulation syndrome (OHSS), enabling preemptive adjustments to treatment plans.

5.8.2 Personalized Approaches to Pregnancy Management

Systems biology also plays a critical role in personalized pregnancy management, especially for high-risk pregnancies. By integrating data from genomics, proteomics, and environmental factors, systems biology models can identify women at risk for complications such as preeclampsia, gestational diabetes, or preterm birth. This allows for early interventions, personalized monitoring plans, and tailored nutritional and lifestyle recommendations, aiming to optimize maternal and fetal health throughout pregnancy.

The importance of systems biology in personalized reproductive medicine has been elaborated upon in Table 5.8.

TABLE 5.8 Role of systems biology in personalizing reproductive medicine, its applications, and upcoming challenges.

Aspect	Application of Systems Biology	Outcomes	Challenges
Personalized Fertility Treatments	Optimizing IVF protocols and embryo selection based on patient data.	Improved conception rates, reduced risk of OHSS.	Need for personalized predictive models and ethical considerations in data management.
Pregnancy Management	Identifying risks and tailoring management plans for high-risk pregnancies.	Early interventions, personalized care to optimize maternal and fetal health.	Integration of complex data into clinical practice, validation of models.
Treatment of Reproductive Disorders	Developing personalized strategies for disorders like PCOS and endometriosis.	Targeted therapies, minimized treatment trial-and-error, enhanced outcomes.	Comprehensive analysis of molecular pathways and genetic factors.
Role of Bioinformatics	Analyzing omics data to guide clinical decisions and predict treatment responses.	Pattern identification, disease risk prediction, effective intervention determination.	Large-scale, longitudinal studies for model validation, advanced computational tools.

5.8.3 Targeted Treatment of Reproductive Disorders

For reproductive disorders like PCOS, endometriosis, and recurrent pregnancy loss, systems biology facilitates the development of personalized treatment strategies. By elucidating the molecular pathways and genetic factors contributing to these conditions, systems biology enables the identification of targeted therapeutic agents that are more likely to be effective for individuals based on their specific molecular profiles. This approach minimizes trial-and-error in treatment selection, reduces adverse effects, and improves overall treatment outcomes.

5.8.4 The Role of Bioinformatics in Personalized Reproductive Health

Bioinformatics is indispensable in personalized medicine, providing the tools needed to analyze and interpret the vast datasets generated by omics studies. Advanced algorithms and machine learning techniques process these data to identify patterns, predict disease risk, and determine the most effective interventions. In reproductive health, bioinformatics supports the identification of genetic variants associated with fertility issues, guides the selection of optimal embryos for implantation, and predicts individual responses to specific drugs or treatments.

While the potential of systems biology in personalized medicine for reproductive health is vast, several challenges remain. These include the need for large-scale, longitudinal studies to validate predictive models, ethical considerations in genetic testing and data management, and the integration of systems biology approaches into routine clinical practice. Future directions will likely focus on enhancing computational models, developing non-invasive diagnostic tools, and implementing patient-centered care models that effectively incorporate personalized medicine strategies.

5.9 CONCLUSION

In conclusion, the exploration of systems biology approaches to reproductive health illuminates a transformative landscape where interdisciplinary collaboration and cutting-edge technologies converge to revolutionize our understanding and management of reproductive disorders. The integration of genomic and proteomic analyses, computational modeling, machine learning, epigenetics, and bioinformatics presents a comprehensive framework for unraveling the complexities inherent in reproductive processes and diseases. At the heart of these approaches lies the promise of precision medicine, offering tailored diagnostics and treatment strategies that account for the individual's genetic makeup, molecular landscape, and environmental context. By deciphering molecular pathways, networks, and epigenetic mechanisms, systems biology empowers researchers and clinicians to identify novel biomarkers, therapeutic targets, and predictive models that cater to the unique needs of each individual facing reproductive challenges.

Moreover, systems biology holds immense potential in enhancing prenatal care, optimizing fertility outcomes, managing pregnancy complications, and mitigating the transmission of reproductive disorders to future generations. Through its holistic view of biological systems, systems biology offers insights into prenatal development, enabling early detection of pregnancy complications and the development of targeted interventions to improve pregnancy outcomes and fetal health. The convergence of computational models with clinical practice represents a pivotal moment in reproductive health care, where theoretical predictions translate into tangible health benefits for patients. As these models become increasingly refined and integrated, the future of reproductive healthcare will be characterized by personalized, effective, and patient-focused interventions that address the diverse needs of individuals across the spectrum of reproductive health.

Furthermore, the ongoing evolution of systems biology in tandem with advancements in machine learning and bioinformatics promises to further revolutionize the field, leading to breakthroughs in diagnostics, therapeutics, and personalized medicine. By harnessing the power of data analytics and computational algorithms, researchers and clinicians can leverage machine learning to enhance diagnostic accuracy, improve understanding of reproductive health issues, and pave the way for personalized treatment strategies tailored to the unique molecular landscape of each individual. In essence, the chapter on systems biology approaches to reproductive health underscores the pivotal role of interdisciplinary collaboration and innovative technologies in shaping the future of reproductive health research and care. As our understanding of the complex biological systems involved in reproduction deepens, the promise of personalized medicine to improve fertility outcomes, manage pregnancy complications, and treat reproductive disorders becomes increasingly tangible.

By continuing to integrate systems biology with clinical practice, the field of reproductive healthcare stands poised to enter a new era of personalized, effective, and patient-focused interventions that prioritize the well-being of individuals and populations alike. In this journey toward advanced healthcare, the importance of collaboration, innovation, and a commitment to bridging the gap between theory and practice cannot be overstated.

REFERENCES

Aderem A (2005) Systems biology: its practice and challenges. *Cell* 121:511–513. https://doi.org/10.1016/j.cell.2005.04.020

Agarwal A, Parekh N, Panner Selvam MK, et al (2019) Male oxidative stress infertility (MOSI): proposed terminology and clinical practice guidelines for management of idiopathic male infertility. *World J Mens Health* 37:296. https://doi.org/10.5534/wjmh.190055

Argelaguet R, Cuomo ASE, Stegle O, Marioni JC (2021) Computational principles and challenges in single-cell data integration. *Nat Biotechnol* 39:1202–1215. https://doi.org/10.1038/s41587-021-00895-7

Bahmyari S, Jamali Z, Khatami SH, et al (2021) microRNAs in female infertility: an overview. *Cell Biochem Funct* 39:955–969. https://doi.org/10.1002/cbf.3671

Bhattacharya A, Freedman AN, Avula V, et al (2022) Placental genomics mediates genetic associations with complex health traits and disease. *Nat Commun* 13:706. https://doi.org/10.1038/s41467-022-28365-x

Buckberry S, Bianco-Miotto T, Bent SJ, et al (2017) Placental transcriptome co-expression analysis reveals conserved regulatory programs across gestation. *BMC Genom* 18:10. https://doi.org/10.1186/s12864-016-3384-9

Calicchio R, Doridot L, Miralles F, et al (2014) DNA methylation: an epigenetic mode of gene expression regulation in reproductive science. *CPD* 20:1726–1750. https://doi.org/10.2174/13816128113199990517

Campbell KA, Colacino JA, Puttabyatappa M, et al (2023) Placental cell type deconvolution reveals that cell proportions drive preeclampsia gene expression differences. *Commun Biol* 6:264. https://doi.org/10.1038/s42003-023-04623-6

Cariati F, D'Argenio V, Tomaiuolo R (2019) The evolving role of genetic tests in reproductive medicine. *J Transl Med* 17:267. https://doi.org/10.1186/s12967-019-2019-8

Casarini L, Crépieux P (2019) Molecular mechanisms of action of FSH. *Front Endocrinol* 10:305. https://doi.org/10.3389/fendo.2019.00305

Clément F (2016) Multiscale mathematical modeling of the hypothalamo-pituitary-gonadal axis. *Theriogenology* 86:11–21. https://doi.org/10.1016/j.theriogenology.2016.04.063

Collins FS, Varmus H (2015) A new initiative on precision medicine. *N Engl J Med* 372:793–795. https://doi.org/10.1056/NEJMp1500523

Curchoe CL, Malmsten J, Bormann C, et al (2020) Predictive modeling in reproductive medicine: where will the future of artificial intelligence research take us? *Fertil Steril* 114:934–940. https://doi.org/10.1016/j.fertnstert.2020.10.040

Day FR, Ruth KS, Thompson DJ, et al (2015) Large-scale genomic analyses link reproductive aging to hypothalamic signaling, breast cancer susceptibility and BRCA1-mediated DNA repair. *Nat Genet* 47:1294–1303. https://doi.org/10.1038/ng.3412

Donkin I, Barrès R (2018) Sperm epigenetics and influence of environmental factors. *Mol Metabol* 14:1–11. https://doi.org/10.1016/j.molmet.2018.02.006

Dumesic DA, Oberfield SE, Stener-Victorin E, et al (2015) Scientific statement on the diagnostic criteria, epidemiology, pathophysiology, and molecular genetics of polycystic ovary syndrome. *Endocrine Rev* 36:487–525. https://doi.org/10.1210/er.2015-1018

Dunaif A, Fauser BCJM (2013) Renaming PCOS—a two-state solution. *J Clin Endocrinol Metabol* 98:4325–4328. https://doi.org/10.1210/jc.2013-2040

Elmannai H, El-Rashidy N, Mashal I, et al (2023) Polycystic ovary syndrome tetection machine learning model based on optimized feature selection and explainable artificial intelligence. *Diagnostics* 13:1506. https://doi.org/10.3390/diagnostics13081506

Escobar-Morreale HF (2018) Polycystic ovary syndrome: definition, aetiology, diagnosis and treatment. *Nat Rev Endocrinol* 14:270–284. https://doi.org/10.1038/nrendo.2018.24

Feil R, Fraga MF (2012) Epigenetics and the environment: emerging patterns and implications. *Nat Rev Genet* 13:97–109. https://doi.org/10.1038/nrg3142

Feinberg AP, Koldobskiy MA, Göndör A (2016) Epigenetic modulators, modifiers and mediators in cancer aetiology and progression. *Nat Rev Genet* 17:284–299. https://doi.org/10.1038/nrg.2016.13

Fröhlich H, Balling R, Beerenwinkel N, et al (2018) From hype to reality: data science enabling personalized medicine. *BMC Med* 16:150. https://doi.org/10.1186/s12916-018-1122-7

Fu K, Li Y, Lv H, et al (2022) Development of a model predicting the outcome of in vitro fertilization cycles by a robust decision tree method. *Front Endocrinol* 13:877518. https://doi.org/10.3389/fendo.2022.877518

Giacone F, Cannarella R, Mongioì LM, et al (2019) Epigenetics of male fertility: effects on assisted reproductive techniques. *World J Mens Health* 37:148. https://doi.org/10.5534/wjmh.180071

Giudice LC, Oskotsky TT, Falako S, et al (2023) Endometriosis in the era of precision medicine and impact on sexual and reproductive health across the lifespan and in diverse populations. *FASEB J* 37:e23130. https://doi.org/10.1096/fj.202300907

Gómez-Roig MD, Pascal R, Cahuana MJ, et al (2021) Environmental exposure during pregnancy: influence on prenatal development and early life: a comprehensive review. *Fetal Diagn Ther* 48:245–257. https://doi.org/10.1159/000514884

Hood L, Flores M (2012) A personal view on systems medicine and the emergence of proactive P4 medicine: predictive, preventive, personalized and participatory. *New Biotechnology* 29:613–624. https://doi.org/10.1016/j.nbt.2012.03.004

Hood L, Galas D (2003) The digital code of DNA. *Nature* 421:444–448. https://doi.org/10.1038/nature01410

Hopkins AL (2008) Network pharmacology: the next paradigm in drug discovery. *Nat Chem Biol* 4:682–690. https://doi.org/10.1038/nchembio.118

Horgan RP, Kenny LC (2011) 'Omic' technologies: genomics, transcriptomics, proteomics and metabolomics. *Obst & Gynaecol* 13:189–195. https://doi.org/10.1576/toag.13.3.189.27672

Huntriss J, Balen A, Sinclair K, et al (2018) Epigenetics and reproductive medicine: scientific impact paper no. 57. *BJOG* 125:e43–e54. https://doi.org/10.1111/1471-0528.15240

Imanian B, Donaghy J, Jackson T, et al (2022) The power, potential, benefits, and challenges of implementing high-throughput sequencing in food safety systems. *NPJ Sci Food* 6:35. https://doi.org/10.1038/s41538-022-00150-6

Itze-Mayrhofer C, Brem G (2020) Quantitative proteomic strategies to study reproduction in farm animals: female reproductive fluids. *J Proteom* 225:103884. https://doi.org/10.1016/j.jprot.2020.103884

Jirtle RL, Skinner MK (2007) Environmental epigenomics and disease susceptibility. *Nat Rev Genet* 8:253–262. https://doi.org/10.1038/nrg2045

Joyce AR, Palsson BØ (2006) The model organism as a system: integrating "omics" data sets. *Nat Rev Mol Cell Biol* 7:198–210. https://doi.org/10.1038/nrm1857

Karczewski KJ, Snyder MP (2018) Integrative omics for health and disease. *Nat Rev Genet* 19:299–310. https://doi.org/10.1038/nrg.2018.4

Kenny LC, Broadhurst DI, Dunn W, et al (2010) Robust early pregnancy prediction of later preeclampsia using metabolomic biomarkers. *Hypertension* 56:741–749. https://doi.org/10.1161/HYPERTENSIONAHA.110.157297

Kitano H (2002) Computational systems biology. *Nature* 420:206–210. https://doi.org/10.1038/nature01254

Kobayashi H, Sato A, Otsu E, et al (2007) Aberrant DNA methylation of imprinted loci in sperm from oligospermic patients. *Hum Mol Genet* 16:2542–2551. https://doi.org/10.1093/hmg/ddm187

Leng D, Zheng L, Wen Y, et al (2022) A benchmark study of deep learning-based multi-omics data fusion methods for cancer. *Genome Biol* 23:171. https://doi.org/10.1186/s13059-022-02739-2

Levine RJ, Lam C, Qian C, et al (2006) Soluble endoglin and other circulating antiangiogenic factors in preeclampsia. *N Engl J Med* 355:992–1005. https://doi.org/10.1056/NEJMoa055352

Li Y (2018) Epigenetic mechanisms link maternal diets and gut microbiome to obesity in the offspring. *Front Genet* 9:342. https://doi.org/10.3389/fgene.2018.00342

Lowe WL, Karban J (2014) Genetics, genomics and metabolomics: new insights into maternal metabolism during pregnancy. *Diabet Med* 31:254–262. https://doi.org/10.1111/dme.12352

McCarthy MI, Abecasis GR, Cardon LR, et al (2008) Genome-wide association studies for complex traits: consensus, uncertainty and challenges. *Nat Rev Genet* 9:356–369. https://doi.org/10.1038/nrg2344

Mennickent D, Rodríguez A, Opazo MaC, et al (2023) Machine learning applied in maternal and fetal health: a narrative review focused on pregnancy diseases and complications. *Front Endocrinol* 14:1130139. https://doi.org/10.3389/fendo.2023.1130139

Morgan HD, Santos F, Green K, et al (2005) Epigenetic reprogramming in mammals. *Hum Mol Genet* 14:R47–R58. https://doi.org/10.1093/hmg/ddi114

Nagaoka SI, Hassold TJ, Hunt PA (2012) Human aneuploidy: mechanisms and new insights into an age-old problem. *Nat Rev Genet* 13:493–504. https://doi.org/10.1038/nrg3245

Nautiyal H, Imam SS, Alshehri S, et al (2022) Polycystic ovarian syndrome: a complex disease with a genetics approach. *Biomedicines* 10:540. https://doi.org/10.3390/biomedicines10030540

O'Hara L, Livigni A, Theo T, et al (2016) Modelling the structure and dynamics of biological pathways. *PLoS Biol* 14:e1002530. https://doi.org/10.1371/journal.pbio.1002530

Ozen M, Aghaeepour N, Marić I, et al (2023) Omics approaches: interactions at the maternal–fetal interface and origins of child health and disease. *Pediatr Res* 93:366–375. https://doi.org/10.1038/s41390-022-02335-x

Padula AM, Monk C, et al on behalf of program collaborators for Environmental influences on Child Health Outcomes (2020) A review of maternal prenatal exposures to environmental chemicals and psychosocial stressors—implications for research on perinatal outcomes in the ECHO program. *J Perinatol* 40:10–24. https://doi.org/10.1038/s41372-019-0510-y

Pammi M, Aghaeepour N, Neu J (2023) Multiomics, artificial intelligence, and precision medicine in perinatology. *Pediatr Res* 93:308–315. https://doi.org/10.1038/s41390-022-02181-x

Rando OJ, Simmons RA (2015) I'm eating for two: parental dietary effects on offspring metabolism. *Cell* 161:93–105. https://doi.org/10.1016/j.cell.2015.02.021

Santos F, Dean W (2004) Epigenetic reprogramming during early development in mammals. *Reproduction* 127:643–651. https://doi.org/10.1530/rep.1.00221

Sapkota Y, Steinthorsdottir V, Morris AP, et al (2017) Meta-analysis identifies five novel loci associated with endometriosis highlighting key genes involved in hormone metabolism. *Nat Commun* 8:15539. https://doi.org/10.1038/ncomms15539

Shendure J, Ji H (2008) Next-generation DNA sequencing. *Nat Biotechnol* 26:1135–1145. https://doi.org/10.1038/nbt1486

Stepto NK, Moreno-Asso A, McIlvenna LC, et al (2019) Molecular mechanisms of insulin resistance in polycystic ovary syndrome: unraveling the conundrum in skeletal muscle? *J Clin Endocrinol Metabol* 104:5372–5381. https://doi.org/10.1210/jc.2019-00167

Stuppia L, Franzago M, Ballerini P, et al (2015) Epigenetics and male reproduction: the consequences of paternal lifestyle on fertility, embryo development, and children lifetime health. *Clin Epigenet* 7:120. https://doi.org/10.1186/s13148-015-0155-4

Suhorutshenko M, Kukushkina V, Velthut-Meikas A, et al (2018) Endometrial receptivity revisited: endometrial transcriptome adjusted for tissue cellular heterogeneity. *Hum Reprod* 33:2074–2086. https://doi.org/10.1093/humrep/dey301

Urteaga I, Albers DJ, Wheeler MV, et al (2017) Towards personalized modeling of the female hormonal cycle: experiments with mechanistic models and gaussian processes. arXiv:1712.00117 [stat.ML]. https://doi.org/10.48550/ARXIV.1712.00117

Wang K, Li Y, Chen Y (2023) Androgen excess: a hallmark of polycystic ovary syndrome. *Front Endocrinol* 14:1273542. https://doi.org/10.3389/fendo.2023.1273542

CHAPTER 6

Machine Learning Algorithms in Reproductive Health

Ananya Verma, Rajshri Singh, and Sagar Barage

6.1 INTRODUCTION

Machine learning (ML) algorithms are computational procedures that enable a system to collect and integrate information by analyzing large volumes of data. These algorithms improve and broaden the system's abilities by acquiring new knowledge through learning rather than solely relying on pre-programmed instructions (Choudhary and Gianey, 2017). These algorithms may be categorized as Unsupervised, Semi-Supervised, Supervised, and Reinforcement Learning, depending on the training data they use and the learning technique they apply. Supervised learning algorithms are taught by using labeled data, which consists of input and output pairs. Semi-supervised learning methods use a mixture of labeled and unlabeled input to acquire patterns and generate predictions. Unsupervised learning methods, in contrast, do not depend on labeled data and instead concentrate on identifying patterns or structures in the incoming data. Finally, reinforcement learning systems acquire knowledge by engaging in a process of trial and error, whereby they get feedback in the form of rewards or penalties based on their interactions with the environment. The extensive use of ML across several fields is evidenced in its ability to improve and automate a broad range of jobs. as depicted in Figure 6.1.

The objective of this book chapter is to provide a general overview of the various types of ML algorithms and their practical applications in addressing real-world reproductive health concerns.

DOI: 10.1201/9781003487548-6

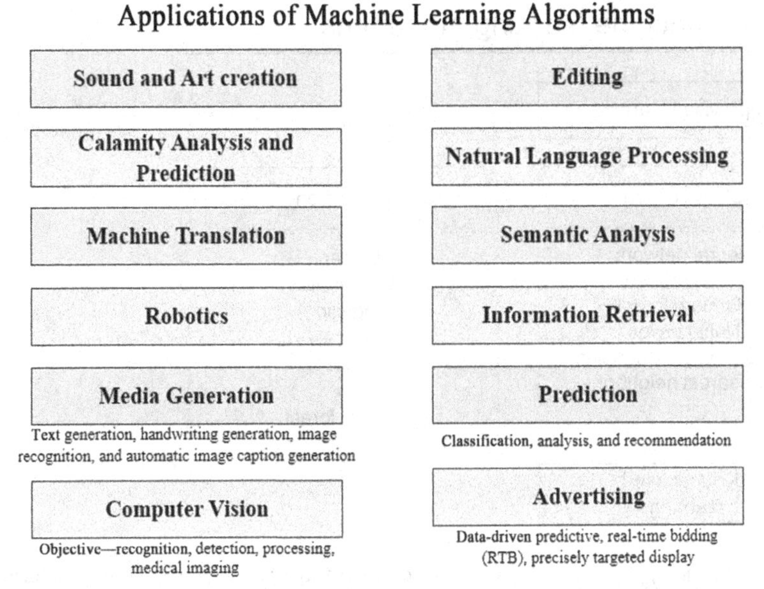

FIGURE 6.1 Applications of ML Algorithm.

6.2 SUPERVISED LEARNING

- Involves the acquisition of knowledge via the use of labeled data, which consists of input–output pairs.
- Establishes a correlation between input features and output labels using training data.
- Estimates the underlying connection between the input characteristics and the output labels.
- This enables the algorithm to make precise predictions on new data.

6.2.1 Classification

- Capable of categorizing data into distinct groups or categories depending on their specific features or attributes from a set of example observations.
- Educate a decision function (classifier) that effectively assigns class labels to new and unseen observations by using the existing patterns.
- ML approaches, such as decision trees, neural networks, naïve Bayes, random forests, k-nearest neighbors (KNNs), and support vector machines (SVMs), may be used for doing this task.

6.2.1.1 Neural Network

- Artificial neural networks (ANNs), also referred to as connectionist models, neurocomputers, or parallel distributed processing models, work similarly to the nervous system of living organisms.

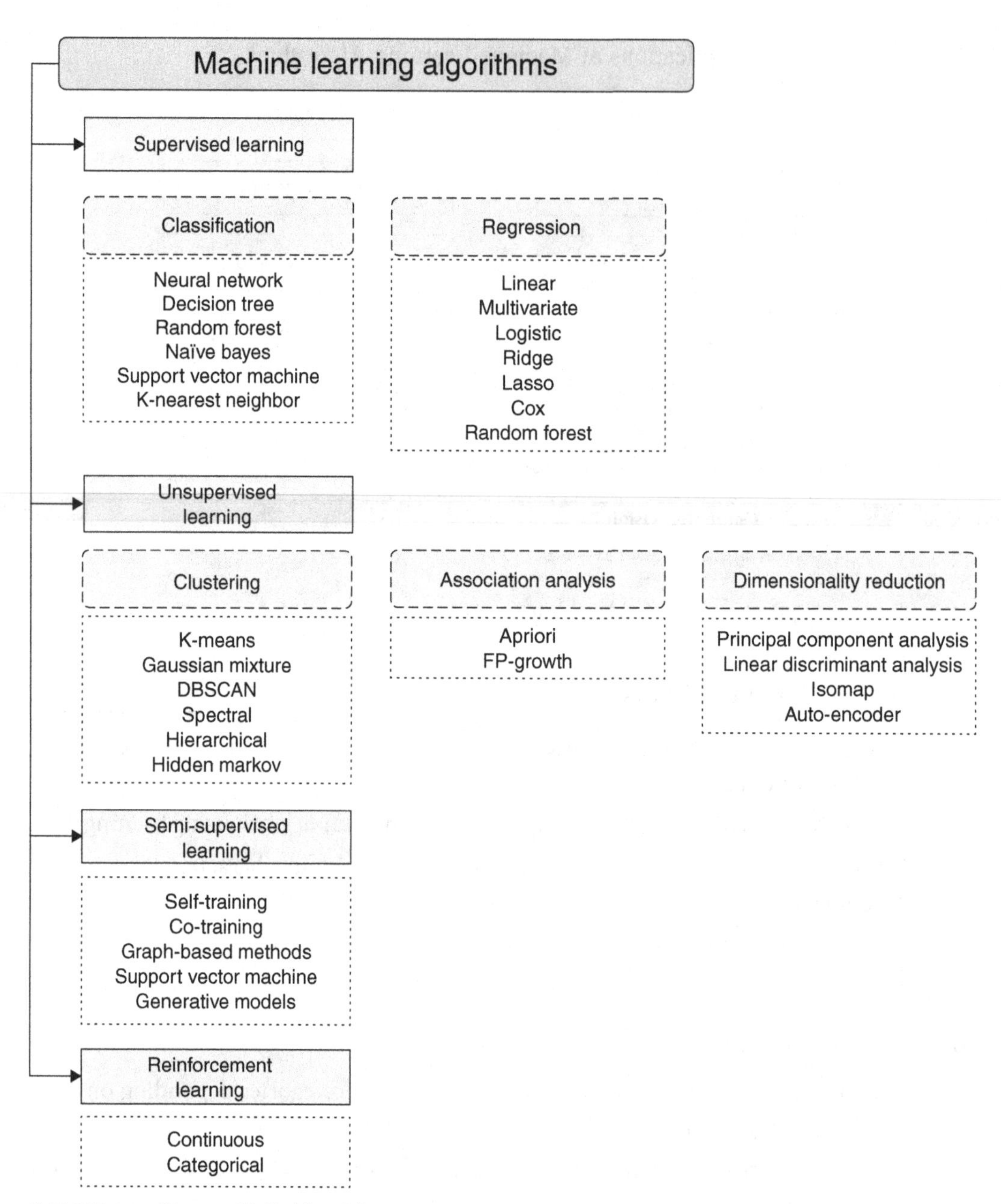

FIGURE 6.2 Types of ML Algorithms.

- Consists of several neurons (or nodes) that operate simultaneously to serve as linked processing units that dynamically adjust the connection strengths between each other to react to external inputs, similar to what occurs during a "stimulus."
- However, the processing result is not saved nor outputted to a designated memory address.
- The overall state of the network is represented after it reaches the state of equilibrium.

- Storage of information is determined by the configuration of the processing units and the synaptic strengths of interneurons and obtained via a process of learning.

6.2.1.2 Decision Tree

- Operates by iteratively dividing the dataset into smaller subsets, taking into account the values of input features.
- Can be used for regression or classification problems, with the main objective of constructing a model that can accurately predict the value of the target variable.
- Learns straightforward decision rules from the input features.
- Ease of understanding, simple methodology, and capacity to handle both categorical and numerical data.
- However, when the trees are deep, they are more likely to overfit the data.
- This issue may be addressed by pruning or using ensemble approaches.

6.2.1.3 Random Forest

- Ensemble approach that consists of many decision trees trained separately on random subsets of data.
- Based on the principle of "wisdom of crowds" – It is made up of multiple trees, and each tree (base learner) forecasts its categorization.
- The ultimate judgment is determined by a model using the largest number of votes from the trees.
- Success lies in the collective strength of uncorrelated trees, as their combined decision-making ability mitigates individual errors, with less correlation among models being crucial for optimal performance.

6.2.1.4 Naïve Bayes

- Based on Bayes' theorem, a mathematical formula is used for conditional probability.
- Assumes that the attributes used to describe an observation are conditionally independent, depending on the class label.
- This is a simplifying assumption, and in reality, features may be correlated, but the algorithm still tends to perform well in practice.

6.2.1.5 Support Vector Machine

- A mathematical method used to optimize a certain mathematical function based on a set of data (sometimes referred to as regularization), which is grounded on the Vapnik–Chervonenkis theory.
- Used to do classification and regression analysis.
- Involves identifying the hyperplane that most effectively divides the data into several groups.

6.2.1.6 K-Nearest Neighbor

- Provided that the class was previously known, the nearest neighbor rule determines the category of an unknown data point by considering its closest neighbor.
- Assumes that data points with similar characteristics are located near each other in the feature space.
- Uses a similarity metric, such as Euclidean distance or cosine similarity, to identify the KNNs of a given data point.
- The class that occurs most frequently among these neighbors is assigned to the new data point.

6.2.2 Regression

- Statistical techniques are used to establish the connection between a dependent variable and one or more independent variables, known as result and predictors, respectively.
- Comprehends how the fluctuation in independent variables alters the value of the dependent variable (Maulud and Abdulazeez, 2020).
- It is often used to forecast or approximate the dependent variable's value based on that of the independent variables.

6.2.2.1 Linear Regression

- Statistical method used to predict the value of a dependent variable by considering one or more independent variables.
- Involves estimating the coefficients of a linear equation to minimize the differences between the predicted and actual output values.
- Often done using "least squares" methods, which assist in identifying the best-fit line for paired data.
- Allows for the determination of the dependent variable based on the independent variable.

6.2.2.2 Multivariate Regression

- An advanced statistical technique that builds upon simple linear regression.
- Allows for the analysis of the intricate relationship between one dependent variable and multiple independent variables.
- Enables the simultaneous investigation of the impacts of several predictors while accounting for the influence of other variables.

6.2.2.3 Logistic Regression

- Statistical method effective for binary classification problems.
- Utilizes regularization to prevent overfitting and enhance model accuracy.

- Employing the logistic function to map input features to probabilities between 0 and 1.
- Widely applied in diverse fields for tasks such as disease diagnosis due to its interpretability and ability to provide probability-based predictions.

6.2.2.4 Ridge Regression

- Also known as L2 regularization, introduces a penalty factor that is directly proportional to the square of the coefficients' magnitude in the least squares objective function.
- Applies a regularization technique that reduces the coefficients toward zero, without setting them precisely to zero.
- Helps decrease the variability of the model and enhances its ability to generalize, particularly when faced with multicollinearity (strong correlation across predictors).

6.2.2.5 Least Absolute Shrinkage and Selection Operator (Lasso) Regression

- Also known as L1 regularization, introduces a component of penalty that is proportional to the true value of the coefficients in the least squares objective function.
- Offers the unique ability to do variable selection by setting certain coefficients to zero.
- Can automatically choose the most significant characteristics.
- Efficiently carries out the process of selecting features inside datasets that include a large number of irrelevant or duplicate features.

6.2.2.6 Cox Regression

- Statistical methods prevalent in survival analysis are integrated into the ML framework through Python libraries like scikit-survival or the R-package called survival.
- Explores the relationship between individuals' survival times and predictor variables, offering functionalities for data handling, model evaluation, feature selection, and evaluation using metrics like concordance index or log-rank test and hyperparameter tuning.

6.2.2.7 Random Forest Regression

- Utilizes the Random Forest ensemble method.
- Employed for regression tasks by constructing multiple decision trees trained on random subsets of data.
- Features at each split point to mitigate overfitting.
- Effectively captures nonlinear relationships in high-dimensional datasets.

6.3 UNSUPERVISED LEARNING

- Discovers patterns and structures from data that do not include any labels or explicit directions in the form of labeled output.
- Analyzes the intrinsic organization of the data to discover noteworthy patterns, connections, or clusters without receiving explicit instructions on what to search for.
- Clustering, Association Analysis, and Dimensionality Reduction are examples of unsupervised learning methodologies.

6.3.1 Clustering

- Groups similar data points based on their attributes or qualities.
- Aims to detect natural clusters or groupings in the data without any preconceived labels.
- Clustering can be conducted in various ways – K-Means, Gaussian Mixture, DBSCAN, Spectral, Hierarchical, and Hidden Markov methods.

6.3.3.1 K-Means

- Divides data points into k clusters by repeatedly allocating every point of data to the closest cluster centroid.
- Revising the centroids to the average of the data points given to each cluster to minimize the sum of squares inside each cluster.

6.3.3.2 Gaussian Mixture

- Probabilistic clustering technique that accommodates the possibility of data points belonging to multiple clusters with varying probabilities.
- Assumes that each of the data points originates from a combination of multiple Gaussian distributions with unidentified parameters.
- Assigns data points to clusters according to the likelihood of them belonging to each cluster.
- Valuable for identifying hidden subgroups within a population.
- Depicts the spatial distribution of a population across extensive regions by categorizing people according to their attributes or trends.
- Enables the implementation of focused treatments and tailored strategies for reproductive health among certain subpopulations.

6.3.3.3 DBSCAN

- "Density-Based Spatial Clustering of Applications with Noise," or DBSCAN, collects data points that are closely packed together in high-density regions.
- Separates outliers into noise points.

- Can handle clusters of arbitrary sizes and shapes.
- Does not require the user to specify the number of clusters in advance.

6.3.3.4 *Spectral Clustering*

- Relies on a graph structure and utilizes the eigenvalues of the similarity matrix to divide the data into distinct groups.
- The process involves using spectral embedding to convert the data into a space with fewer dimensions.
- Followed by the application of K-Means or equivalent clustering methods on the modified data.

6.3.3.5 *Hierarchical Clustering*

- Constructs a hierarchical structure of clusters by iteratively dividing or combining clusters based on their similarity.
- User is not required to predefine the cluster number.
- May generate a dendrogram to see the hierarchical clustering.

6.3.3.6 *Hidden Markov*

- Type of probabilistic graphical model used for modeling sequential data.
- Can be used for clustering sequences of data by identifying hidden states that represent underlying patterns or behaviors in the data.

6.3.2 Association Analysis

- Data mining method that uncovers significant relationships or patterns among variables in extensive datasets.
- Useful for identifying associations or correlations between items in transactional databases.
- Association analysis can be conducted either through Apriori or FP-Growth algorithms.

6.3.2.1 *Apriori*

- Classic algorithm for association rule mining.
- Comprises three main steps – joining of frequent datasets of a particular size to generate candidate datasets of larger size and pruning candidate datasets that violate the Apriori property.

6.3.2.2 *FP-Growth*

- Based on a divide-and-conquer strategy.
- Utilizes the FP-tree data structure and efficiently mines frequent datasets by constructing an FP-tree from the transaction database.

- Traverses it to directly identify frequent datasets via recursive conditional pattern base generation.
- Subsequently, it generates association rules based on a minimum confidence threshold.

6.3.3 Dimensionality Reduction

- Used to reduce the number of features in a dataset while retaining important information.
- Useful in datasets with a large number of dimensions, as it helps address issues such as computational complexity, overfitting, and excessive dimensionality.
- Two main approaches to dimensionality reduction: feature selection and feature extraction, which involves choosing a subset of the original features and transforming the features into a lower-dimensional space, respectively.
- Offers benefits such as enhanced accuracy, but also has drawbacks such as loss of information and reduced interpretability.
- The choice between these approaches depends on factors such as the characteristics of the data and the specific learning tasks at hand (Li et al., 2020).
- There are many strategies for reducing dimensionality, including linear discriminant analysis or LDA, principal component analysis or PCA, Auto-encoder, and Isomap.

6.3.3.1 Principal Component Analysis

- Finds orthogonal axes, known as principal components, to reduce the dimensionality of data.
- Axes represent the highest amount of variation in the data.
- By mapping the data onto principal components and choosing a subset that accounts for a certain proportion of the overall variance, it efficiently decreases the number of dimensions while retaining important information.

6.3.3.2 Linear Discriminant Analysis

- Employed in classification problems with the objective of mapping data into a lower-dimensional subspace while maximizing the distinction between classes.
- Achieved by maximizing the ratio of within-class variance and between-class variance.
- Through the transformation of original features to discriminant functions (linear combinations of the original attributes).
- Facilitates the classification of new data points based on their projections onto these functions.

6.3.3.3 Isomap

- Isomap restores the intrinsic geometrical structure for high-dimensional data in lower-dimensional spaces via Multi-dimensional Scaling (MDS)
- Graphically, it represents data points as nodes and similarities as edges.
- Then, computes the shortest distance between them using specific algorithms or spectral techniques.

6.3.3.4 Auto-encoder

- System that uses ANNs.
- Consists of an encoder network that maps input data to a lower-dimensional latent space and a network of decoders that reassembles the input data from the latent space representation.
- Reduces dimensionality and compresses data.

6.4 SEMI-SUPERVISED LEARNING

- Aims to enhance the accuracy of the learning algorithm by utilizing the extra information present in unlabeled data.
- Useful in situations where acquiring labeled data is costly or time-consuming.
- Encompasses many approaches that may be used to address data-driven challenges, including SVM, Self-training, Graph-based techniques, Co-training, and Generative models.

6.4.1 Self-training

- Generates predictions on labeled data is first trained using the labeled data.
- Predicted labels that have a high level of confidence are then included in the labeled dataset.
- Undergoes retraining using the enhanced labeled dataset, and this process continues until convergence is achieved.

6.4.2 Co-training

- Entails training several models or perspectives on distinct subsets of characteristics or depictions of the data.
- At first, every model is trained using the labeled data.
- Share reliably predicted labels on the unlabeled data with one another.
- The models undergo retraining using the supplemented labeled data, and this procedure is repeated frequently.

6.4.3 Graph-based Methods

- Represents the data instances as nodes in a graph, where edges represent correlations between instances.
- Utilizes the graph structure to propagate label information from labeled to unlabeled nodes through the graph.
- Label propagation algorithms, such as label spreading or label propagation, are commonly used in this approach.

6.4.4 Semi-supervised Generative Models

- Integrates unsupervised and supervised learning by including latent variables to represent the underlying data distribution.
- Generative adversarial networks (GANs) or variational autoencoders (VAEs) are capable of generating realistic data samples and predicting labels for labeled instances concurrently.

6.4.5 Semi-supervised Support Vector Machines

- Semi-supervised SVMs extend traditional SVMs by incorporating unlabeled data to enforce a smoother decision boundary.
- Enhances classification performance by leveraging the margin between decision boundaries depending on the nature of the data.
- Availability of labeled data.
- The specific learning task at hand.

6.5 REINFORCEMENT LEARNING

- Framework where an agent acquires the ability to make choices through repeated environmental interactions.
- The agent is provided with rewards or penalties as feedback, which are determined by the state it transitions into and the activities it does.
- The agent's goal is to acquire a strategy or policy that maximizes the total reward it gets over some time.

6.5.1 Continuous Reinforcement

- Continuous reinforcement learning involves an action space characterized by a continuum of values, allowing actions to span a range within a continuous domain.
- Commonly implemented in tasks where actions are not confined to distinct choices but can instead encompass any value within a specified series.

6.5.2 Categorical Reinforcement

Categorical reinforcement learning pertains to tasks where discrete actions are drawn from a finite set of categories.

Provokes the agent to select from a predefined set of actions, where actions could correspond to specific moves or denote choices of items to recommend.

6.6 REAL-WORLD APPLICATIONS OF MACHINE LEARNING ALGORITHMS IN REPRODUCTIVE HEALTH

6.6.1 Puberty

The study conducted by Pan et al. (2019) employed multiple ML algorithms to create prediction models. These models were designed to address nonlinear interactions between features, determine feature importance, and provide good interpretability. The researchers utilized Extreme Gradient Boosting (XGBoost) due to its effectiveness in training a sequence of models to minimize errors and its scalability. They also used Random Forests to handle nonlinear data and ensure robustness to noise. The study conducted by Pan et al. (2019) used SVMs and Decision Trees to assess the performance of ensemble models vs nonensemble models. This analysis provided useful insights into the connections between input variables and the output gonadotropin-releasing hormone analogue (GnRHa) test findings in girls with presumed central precocious puberty.

6.6.2 Menstrual Cycle Tracking

The study conducted by Yu et al. (2022) revealed that the integration of wearable devices and algorithms for ongoing monitoring of health data might potentially enhance the chances of successful conception in women. The research used a range of ML methodologies, which were implemented and shown using Python. The algorithms were developed using basal body temperature and heart rate data acquired directly from participants. Their performance was assessed based on evaluation metrics and parameters. The objective was to forecast the fertile period and menstruation in individuals with both consistent and inconsistent menstrual cycles. In addition, the researchers (Yu et al., 2022) examined the incorporation of physiological parameters, such as the variability of the low-frequency to high-frequency ratio and normal-to-normal intervals, into predictive models.

6.6.3 Family Planning

Kebede et al. (2023) used a range of ML techniques to identify the main factors contributing to the lack of access to family planning services among women in Ethiopia. The selection of these algorithms was based on their distinct strengths and weaknesses to thoroughly analyze and select the most optimal model for predicting this occurrence. The Random Forest technique has the highest accuracy rate of 85%, with an AUC (area under the curve)] value of 0.93. The study identified many key factors associated with insufficient resources for family planning, including rejection from husbands/partners, women's educational attainment, geographic region, previous childbirth in a healthcare facility, financial position, and household size (Kebede et al., 2023).

6.6.4 Contraceptive Use

ML algorithms and iterative thematic analysis (ITA) were employed to identify the factors influencing the use of intrauterine devices (IUDs) among married women in India. This approach provided valuable insights into the variable constructs derived from a set of regularized models (Dey et al., 2022). The Lasso Logistic Regression method was used to determine the key predictors of IUD usage. This technique reduces the coefficients of extraneous variables to zero, thereby removing unrelated variables from the dataset. This results in a more concise and focused input dataset. Ridge Logistic Regression was used to mitigate multicollinearity among the variables by including a regularization factor in the regression equation. The neural network used variables or characteristics with non-zero coefficient estimates from Lasso as input units, with women's usage of IUD as the output. This was done to consider nonlinear correlations among the predictors. The findings were then compared to those obtained using the iterative theme categorization technique. The algorithms' analysis revealed that several factors strongly influenced the use of IUDs among married women in India. These factors included shared family goals, utilization of family planning services, intention to have or not have children, wealth, education, access to counseling, and availability of maternal and child healthcare services.

6.6.5 Unintended Pregnancy

The researchers used several ML methods to forecast instances of unplanned pregnancies among married women in Bangladesh (Hossain et al., 2022). The Elastic Net Regression algorithm demonstrated superior performance in accurately classifying unintended pregnancies and their subsequent interventions among married couples in Bangladesh, achieving an impressive AUC of 74%. Furthermore, it identified age, education, wealth status, and contraceptive intention as the primary factors influencing the occurrence of these pregnancies.

6.6.6 Assisted Reproductive Technology

The method used to evaluate individual treatment effects for improving trigger injection time for in vitro fertilization (IVF) results was a T-learner with bagged Light Gradient Boosting Machine (LightGBM) base learners. The technique was selected due to its capacity to handle categorical and continuous covariates, capture nonlinear relationships, efficiently process huge datasets, and consider nonlinear interactions between input variables. The analysis included a range of patient characteristics and stimulation settings. The primary objective was to use ML to assist in making crucial decisions during IVF procedures. This might potentially lead to improved clinical outcomes by optimizing the yield of fertilized oocytes (2PNs) and the total number of viable blastocysts obtained from a specific cohort of stimulated follicles. Hariton et al. (2021) showed that using algorithms to aid doctors in making choices during IVF treatment may lead to a rise of

1.430 2PNs and 0.525 total useable blastocysts per cycle, as compared to judgments made without assistance. Furthermore, the research highlighted the capacity of ML algorithms to assist and enhance physician judgments, resulting in cost-efficient and effective progress in IVF therapy.

6.6.7 Maternal Healthcare

The research carried out by Mutlu et al. (2023) highlighted the potential benefits of using ML methods to detect maternal health hazards. It also underlined the need for early identification of risk factors for the welfare of pregnant women. The research conducted a comparison of several ML techniques. The objective was to assess medical variables such as maternal age, body temperature, blood oxygen saturation, blood pressure, heart rate, and their respective measurements. The aim was to assess the degree of risk severity for pregnant women. The research findings indicated that the Decision Tree had the greatest level of accuracy, reaching 89.16%, while assessing the risk of maternal health. On the other hand, among the methods investigated in the study, the KNN approach had the lowest accuracy rate, measuring 68.47% (Mutlu et al., 2023).

Numerous ML paradigms were employed in approximately 127 studies reviewed by Davidson and Boland (2021) that aimed to enhance pregnancy outcomes by focusing on different aspects of pregnancy care, such as prenatal care, perinatal care, birth and delivery, and preterm birth. Even though unsupervised learning methods (n = 9) were found to be less popular than supervised methods (n = 69), deep learning has gained popularity for its ability to provide valuable insights and enhance patient care by optimizing pregnancy outcomes, especially in addressing maternal healthcare needs, postpartum care, and pregnancy care for transgender individuals (Davidson and Boland, 2021).

In their 2020 study, Sufriyana et al. conducted a systematic review and meta-analysis to compare the predictive abilities of different ML algorithms. The purpose of the study was to develop or validate a multivariable prognostic prediction model for pregnancy care. Logistic Regression was the predominant algorithm used for constructing a predictive model in prenatal care. Random Forest and Gradient Boosting algorithms have shown strong performance in predicting certain pregnancy outcomes, including preterm birth, pre-eclampsia, cesarean section, and gestational diabetes.

The prediction model for preterm birth was developed using Electronic health records were used as the basis for the model, and its performance was assessed using several evaluation metrics (Sun et al., 2022). The study involved 9550 pregnant women, divided into two groups: 4775 women who had preterm births and 4775 women who served as controls. The average age of the women in the preterm birth group was lower than that of the control group. Additionally, the gestation period of the women in the preterm birth group was considerably shorter than that of the control group. There were also differences between the two groups in terms of characteristics such as parity, gravidity, newborn height, weight, and Apgar scores. The performance of the prediction model was evaluated using many metrics. It was determined that the Random Forest Tree model had the best accuracy of 0.816 and an AUC-ROC curve (area under the receiver operating characteristic curve) of 0.885 (with a 95% confidence range of 0.873–0.897) compared to the other models.

6.6.8 Predicting Fetal Risk using Cardiotocograph Data

The performance of various ML models was evaluated during the training phase. This evaluation was conducted on a cardiotocograph (CTG) dataset that was divided into training and testing datasets using the K-Fold Cross Validation technique. The XGBoost model demonstrated the maximum accuracy (93%) compared to other ML models in predicting suspect and pathological fetal states based on CTG tracings (Hoodbhoy et al., 2019).

6.6.9 Childbirth

The various modes of childbirth – cesarean birth, emergency cesarean, forceps delivery, vacuum extraction, and vaginal birth; were explored by Islam et al. (2021) via machine algorithms based on a 6157 births' dataset and a specific set of features. The results obtained established that there were modes of childbirth that could be predicted based on 32 features (such as height, age, allergies, body mass index (BMI), weight before pregnancy, previous pregnancies' history, and more), and ML predictive modeling obtained F1-scores of 97.9%, 97.3%, 95.8%, 93.2%, and 88.6%, respectively (Islam et al., 2021).

Predictive analysis was conducted to determine the use of skilled delivery services by pregnant women in Ethiopia. This analysis involved the utilization of the Statistical Package for Social Sciences (SPSS) for logistic regression and the development of models using several classification algorithms in the Waikato Environment for Knowledge Analysis (WEKA). In addition, the sub-models were created utilizing all input variables and derivation datasets, with the primary factors being first prenatal care, birth order, contraceptive usage, television ownership, healthcare costs, age at first birth, and age at first sexual intercourse. Among all the model-building algorithms, J48 exhibited exceptional performance according to traditional predictive metrics, including accuracy (98%), specificity (99%), sensitivity (96%), and the area under the receiver operating characteristic (ROC) curve (98%). The study found that a mere 27.7% of women in Ethiopia had gotten competent delivery support. Additionally, the researchers suggested that ML algorithms may be used to create prediction models that can help focus treatments and assure expert assistance during labor (Tesfaye et al., 2019).

6.7 REPRODUCTIVE HEALTH DISORDERS

Healthcare providers can utilize ML techniques to integrate multi-omics data and develop predictive models for obstetrical complications, including gestational diabetes and pre-eclampsia. They can also use these techniques to monitor mental health issues such as anxiety, depression, and postpartum depression during and after pregnancy. Additionally, ML can aid in biomarker identification to improve the accuracy of disease prediction, stratification of patients, and the delivery of precision or personalized medicine. Furthermore, ML can optimize wearable devices and mobile apps to effectively track overall reproductive well-being (Kharb and Joshi, 2023).

6.7.1 Pre-eclampsia

The study by Li et al. (2021) employed several ML algorithms (Random Forests, Logistic Regression, SVM, and XGBoost), chosen for their suitability in binary classification tasks based on the collected clinical parameters collected at the first antenatal care visit to construct the prediction model for the risk of pre-eclampsia (Li et al., 2021). The study utilized 38 candidate clinical parameters, including both binary (nulliparity, multifetal gestations, pre-gestational diabetes, autoimmune diseases, etc.) and continuous variables (maternal age and BMI).

6.7.2 Endometriosis

In their study, Matta et al. (2020) utilized several ML algorithms. They applied these algorithms to a dataset obtained from a case-control study conducted in France. The main objective of the study was to identify important biomarkers of internal exposure in adipose tissue that are associated with endometriosis. Additionally, the researchers compared the performance of different models to gain a comprehensive and precise understanding of the complex relationships between exposure to combinations of persistent organic pollutants (POPs) and the increased likelihood of developing endometriosis. The study conducted by Matta et al. in 2020 repeatedly found a correlation between deep endometriosis and some POPs such as cis-heptachlor epoxide, octachlorodibenzofuran, trans-nonachlor, and polychlorinated biphenyl.

6.7.3 Cervical Cancer

Three ML algorithms – SVM, XGBoost, and Random Forest – were used to analyze the dataset of cervical cancer; of these, XGBoost and Random Forest performed better than SVM (Deng et al., 2018). It was found that to detect and treat cervical cancer, it is necessary to take into account a potential patient's age, age of first sexual intercourse, their use of hormonal contraceptives, the number of sexual partners they have had as well as the number of times they have gotten pregnant.

6.7.4 Human Immunodeficiency Viruses

The predictive capabilities of several ML algorithms and Deep Learning (DL) were utilized to accurately forecast the likelihood of contracting human immunodeficiency virus (HIV), syphilis, gonorrhea, and chlamydia (Xu et al., 2022). While nonlinear ML methods such as GBM and RF performed better than traditional regression models, it was shown that combining ENR, GBM, and RF in a stacking ensemble methodology yielded even better results than using the individual ML models alone. The top 10 predictors for each of the four illnesses evaluated accounted for eight percent of the total STI and HIV model performance. The external validation findings showed a fairly similar area under the curve, ranging from 0.69 to 0.85, compared to the analysis of the testing data. Moreover, a web-based risk prediction tool was created and verified to assess an individual's susceptibility to HIV and STIs.

In their research, Comulada et al. (2021) employed 10-fold cross-validation to determine the best penalty parameters for ENR and Lasso. They then minimized overfitting and errors by dividing the training data into separate training and validation datasets. Using these datasets, they constructed predictive models based on the assessed characteristics in an HIV prevention trial. Comulada et al. (2021) identified new connections between sociodemographic factors, health risks, and the behavior of seeking health information on the internet. The study also examined factors associated with internet health seeking, such as the use of PrEP/PEP (pre-exposure prophylaxis/post-exposure prophylaxis), previous participation in HIV prevention programs, access to healthcare, minority status, and homelessness.

6.7.5 Klinefelter Syndrome

The prediction of Klinefelter syndrome in prepubertal boys was made using a trained random forest model. This model classified unseen observations based on the SDS (sodium dodecyl sulfate) profile of serum concentrations of LHluteinizing hormone), inhibin B, FSH follicle-stimulating hormone), AMH anti-Müllerian hormone), total testosterone, SHBG, 17-OHP, A4, DHEAS (dehydroepiandrosterone sulfate), and body height. The accuracy of the model was evaluated using a confusion matrix, which showed an accuracy of 77.8% and an average AUC-ROC of 82.4%. These results indicate that the model performed better than other models mentioned in the study by Madsen et al. (2023). The findings indicated that prepubertal boys with Klinefelter syndrome had significantly reduced levels of LH, FSH, DHEAS, and 17-OHP in contrast to age-matched prepubertal controls. However, they displayed notably elevated levels of AMH.

Among the ML models tested, SVM and Gradient Boosting on Decision Trees (CatBoost) were identified as the most suitable models for distinguishing between Klinefelter Syndrome cases and non-Klinefelter cases. These models achieved a sensitivity of 100% and a specificity of over 93% on the test dataset. Furthermore, they outperformed a group of 18 expert clinicians who achieved a sensitivity of 87% and a specificity of 89.9%. This finding was reported by Krenz et al. in 2022. Furthermore, a Klinefelter Syndrome Score Calculator was devised using predictive models.

6.7.6 Polycystic Ovary Syndrome

Polycystic ovary syndrome (PCOS) is a prevalent hormonal disorder among women of child-bearing age, with significant implications for reproductive health, including anovulation, infertility, cardiovascular disease, obesity, and type 2 diabetes. Numerous studies have been conducted for the early detection and prevention of PCOS, including those carried out by Denny et al. (2019) and Thakre (2020). Each systematic comparison was made between ML algorithms, and Random Forest demonstrated the best performance every time.

Prediction of PCOS using clinical and metabolic markers was carried out by first transforming the dataset using PCA and then subjecting it to ML algorithms in the Spyder Python IDE (Denny et al., 2019). With an accuracy rate of 89.02%, the Random Forest Classifier (RFC) proved to be the most appropriate and precise technique for predicting PCOS.

Thakre (2020) used a dataset containing 41 features, of which 30 were considered as prominent using the CHI SQUARE method used in the feature vector. The authors compared the results of each classifier and found that the accuracy of the RFC (90.9%) was the highest and the most reliable. Furthermore, the publication also included the K Fold cross-validation scores and the accuracy, recall, and F-score for each model on the testing data, which served to enhance the credibility of the evaluation (Thakre, 2020).

6.7.7 Female Factor Infertility

The Naïve Bayes' Classifier, C4.5 decision trees algorithm, and Multi-layer Perceptron (MLP) were implemented using the WEKA environment. They were evaluated using the 10-fold cross-validation method to predict infertility likelihood in women and derive a general principle for analyzing patient risk factor data to forecast the chance of infertility. The MLP and C4.5 algorithms demonstrated superior performance compared to the Naïve Bayes' Classifier, achieving an accuracy of 77.4% (Balogun et al., 2018).

6.7.8 Male Factor Infertility

The models Random Forest, Stochastic Gradient Boosting, Lasso, Ridge Regression, and XGBoost were trained on a dataset that was randomly divided into 80% for training and 20% for testing. The models were then evaluated using four key evaluation metrics: relative absolute error, symmetric mean absolute percentage error, root relative squared error and root mean squared error. The best-performing model for each of the models was selected based on hyperparameter tuning and validation using a 10-fold cross-validation method. The goal was to construct a predictive model for identifying important risk factors affecting sperm count, as described by Huang et al. (2023). The top factors contributing to sperm count, based on results obtained by the five ML paradigms, were ranked respectively as alpha-fetoprotein, sleep time, body fat, blood urea nitrogen, systolic blood pressure, body mass index, uric acid, cholesterol to high-density lipoprotein ratio, total cholesterol, and waist–hip ratio.

Akinsal et al. (2018) examined the use of ANNs in identifying azoospermic patients who require further genetic evaluation based on factors such as total testicular volume and LH levels, in interpreting sperm morphology and predicting semen quality, as well as in recognizing biomarkers for infertility (Akinsal et al., 2018).

6.8 CHRONO-DISRUPTION IN BIOMARKERS

The research by Rúa et al. (2023) used the amalgamation of outcomes of decision tree-based methods – CART trees (rpart and rpart2) and Random Forest (randomForest) algorithm through R programming language and consensus criteria, to mitigate biases for small sample size for the construction of classification models and feature selection to uncover biomarkers associated with the conditions that disturb the body's internal clock and examine their impact on reproductive health. Eating patterns, social jet lag, sleep quality, and sleep duration affect the reproductive health and mental health of the expectant mother during and after pregnancy (Rúa et al., 2023).

6.9 INFANT MORTALITY

Several ML algorithms were employed to analyze the Rwanda Demographic and Health Survey 2014–15 dataset and develop predictive models for infant mortality (Mfateneza et al., 2022). Random Forest model demonstrated the best performance in predicting infant mortality, with an accuracy of 84.7% and an AUC-ROC of 83%, followed by the

SVM showing an accuracy of 68.6%, thus outperforming traditional logistic regression models.

6.10 TOOLS

6.10.1 DeepFert

The research by Naseem et al. (2023) utilized several ML algorithms and Deep Learning-based Neural Networks (DLNN) to compare their effectiveness in predicting semen quality and fertility rates as well as propose a method for use in automated artificial insemination workflows. DLNN achieved comparatively significant results with a semen prediction accuracy of 80.952% and a sperm concentration accuracy of 85.714% with great effectiveness and consistency (Naseem et al., 2023).

6.10.2 MotilitAI

The motilitAI framework utilizes multiple ML algorithms, including Linear Support Vector Regressor (SVR), Multilayer Perceptron (MLP), Bag-of-Words with Support Vector Regressor (BoW with SVR), and more. These algorithms are trained on features extracted from sperm videos, specifically displacement vectors and statistics computed from sperm tracks. The performance of the framework is evaluated using metrics such as mean absolute error and root-mean-square error. The objective of this framework is to predict human sperm motility (Ottl et al., 2022). The MLP model achieved the best results when trained on EMSD feature vectors extracted using Trackpy and unsupervised tracking of sperm cells with the Crocker-Grier algorithm. The extracted features were then aggregated into a histogram for each detected track using BoW.

6.10.3 CASAnova

The computer-aided sperm analysis (CASA) nova machine model, which is facilitated by SVMs, can be employed for the determination of motility percentage and kinematic parameters (Goodson et al., 2017). It uses kinematic gating techniques for distinguishing vigorous sperms as progressive, intermediate, or hyperactivated and non-vigorous sperms as slow or weakly motile. With an inclusive accuracy of about 89.9%, it has the capability of allowing the quick quantitative analysis of motility changes within more than 26,000 human sperms during capacitation.

6.10.4 i-HOPE

The system was created by gathering data from numerous clinics and hospitals, which included physiological and metabolic indicators associated with PCOS and infertility. This data was then analyzed using data mining techniques, ultrasound report categorization, and competitive neural networks. ML techniques were employed to classify PCOS. The i-HOPE project offers the potential to explore novel approaches in the treatment of PCOS by investigating the effects of Vitamin D, preterm labor/abortions, and lean body composition on PCOS patients (Denny et al., 2019).

6.11 REPRODUCTIVE HEALTH VULNERABILITY

ML algorithms were used to predict vulnerability based on migrant features and migrating cities in a rule-based, manually curated, and artificial dataset (as there was no readily available dataset for the experiments), with a high recall for the vulnerable class and minimized false negatives (Nigam et al., 2019). Feature importance was evaluated using Random Forest, with "age of migrant" and "accompanying adult in the family" identified as the top predictive features for the dataset based on metrics such as F1 score, accuracy, and confusion matrices. Furthermore, to gather data, provide safety and awareness, and present insights and recommendations to the public as well as professionals involved in migration, including migrants, public health workers, and policymakers, a web app was designed.

6.12 NON-MARITAL SEXUAL VIOLENCE

The research undertaken by Raj et al. (2021) used ML methods to ascertain the parameters linked to Non-Marital Sexual Violence (NMSV) in India. The Lasso and Ridge models were employed, accompanied by the ITA of the generated variables. This analysis helped identify previously unknown factors linked to NMSV, including limited knowledge and utilization of sexual and reproductive health (SRH) services, as well as cultural norms and preferences indicating more rigid gender norms. The Neural Network models were employed to investigate previously unidentified Non-Mutually Exclusive Sexual Violence connections, as well as to reveal the variables associated with violent encounters, exposure, knowledge, and availability of SRH services (Raj et al., 2021).

6.13 DIGITAL HEALTH PROGRAMS

Logistic regression, LDA, SVM, Classification and regression trees, Naïve Bayes, Neural Networks, K-means clustering, and algorithms were used to examine two digital health initiatives – Kilkari, a mobile messaging service designed for pregnant women, and Mobile Academy, a mobile training tool for frontline healthcare workers, to understand several facets of the programs, including user attributes, exposure to message content, and patterns of training commencement and conclusion. The ultimate objective of the study was to assess the predictions of results and their efficacy while also providing useful insights on strategic data collection (Mohan et al., 2019).

Applications of ML Algorithms in Reproductive Health are summarized in Table 6.1.

6.14 ETHICAL CONSIDERATIONS

With the growing utilization and endorsement of ML techniques, it is essential to implement a systematic approach to identify and address the ethical implications of ML in reproductive healthcare (Figure 6.3). Ensuring fairness and minimizing bias in ML algorithms is a critical ethical consideration, many instances of bias can be seen in ML algorithms that are used in the healthcare sector, such as the use of scores that are adjusted based on race and ethnicity, thus interfering with the prioritization and medical interventions, and worsen existing health inequalities. The results generated by ML paradigms are very

TABLE 6.1 Applications of ML Algorithms in Reproductive Health.

S. No.	Application	Algorithms used	Author(s)
1.	Puberty	XGBoost, RF, SVM, DT	Pan et al. (2019)
2.	Menstrual Cycle Tracking	LR, RF, SVM, GBM, ANN	Yu et al. (2022)
3.	Family Planning	LR, RF, KNN, ANN, SVM, NB, XGBoost, AdaBoost	Kebede et al. (2023)
4.	Contraceptive Use	Lasso, Ridge, ANN	Dey et al. (2022)
5.	Unintended Pregnancy	ENR, LR, RF, SVM, KNN	Hossain et al. (2022)
6.	Assisted Reproductive Technology (ART)	LightGBMKNN	Hariton et al. (2021), Fanton et al. (2022)
7.	Maternal Healthcare	DT, LightGBM, CatBoost, RF, GBM, KNNSVM, ANN, RA, RF, DLNNLR, ANN, SVM, DLNN, DT, RF, GBMNB, SVM, RF, ANN, KM LR	Mutlu et al. (2023), Davidson and Boland (2021), Sufriyana et al. (2020), Sun et al. (2022)
8.	Childbirth	SC, RF, KNN, DT, SVMJ48, ANN, NB, SVM	Islam et al. (2021), Tesfaye et al. (2019)
9.	Reproductive Health Disorders		
	• Pre-eclampsia	LR, RF, SVM, XGBoost	Li et al. (2021)
	• Endometriosis	LR, ANN, SVM, AdaBoost, PLSDA	Matta et al. (2020)
	• Cervical Cancer	SVM, XGBoost, RF	Deng et al. (2018)
	• Human Immunodeficiency Viruses (HIV)	LR, Ridge, Lasso, ENR, GBM, RF, NB, DLNNENR, Lasso	Xu et al. (2022), Comulada et al. (2021)
	• Klinefelter Syndrome – Klinefelter Syndrome Score Calculator	RFAdaBoost, GP, KNN, MLP-SKLearn, SVM, CatBoost, MLP-TensorFlow	Madsen et al. (2023), Krenz et al. (2022)
	• Polycystic Ovary Syndrome (PCOS)	NB, LR, DT, RF, SVM, KNN, CART	Denny et al. (2019), Thakre (2020)
	• Female Factor Infertility	NB, C4.5 DT, MLP	Ademola Balogun et al. (2018)
	• Male Factor Infertility	RF, SGB, Lasso, Ridge, XGBoostANN	Huang et al. (2023), Akinsal et al. (2018)
10.	Chrono-disruption in biomarkers	CART, RF	Rúa et al. (2023)
11.	Infant Mortality	RF, DT, SVM, LR	Mfateneza et al. (2022)
12.	Tools		
	• DeepFert	RF, SVM, MLP, DT, DLNN	Naseem et al. (2023)
	• MotilitAI	SVR, MLP, BoW with SVR, BoW with MLP, CNNs	Ottl et al. (2022)
	• CASAnova	SVM Regression	Goodson et al. (2017)
	• i-HOPE	LR, LDA, KNN, CART, RF, NB, SVM	Denny et al. (2019)
13.	Reproductive Health Vulnerability	RF, SVM, XGBoost, SNN	Nigam et al. (2019)

TABLE 6.1 (Continued)

S. No.	Application	Algorithms used	Author(s)
14.	Non-Marital Sexual Violence (NMSV)	Lasso, Ridge, ANN	Raj et al. (2021)
15.	Monitoring and Evaluating Digital Health Program – Kilkari, Mobile Academy	LR, LDA, SVM, CART, Naïve Bayes, Neural Networks (NNs), K-means clustering	Mohan et al. (2019)

Abbreviations: AdaBoost – Adaptive Boosting, **ANN** – Artificial Neural Networks, **BoW** – Bag-of-Words, **CART** – Classification and Regression Trees, **CNN** – Convolutional Neural Networks, **DT** – Decision Trees, **DLNN** – Deep Learning-Based Neural Network, **ENR** – Elastic Net Regression, **XGBoost** – Extreme Gradient Boosting, **GP** – Gaussian Process, **GBM** – Gradient Boosting Machine, **CatBoost** – Gradient Boosting on Decision Trees, **KM** – K-Means, **KNN** – K- Nearest Neighbor, **Lasso** – Lasso Regression, **LightGBM** – Light Gradient Boosting Machine, **LDA** – Linear Discriminant Analysis, **SVR** – Linear Support Vector Regressor, **LR** – Logistic Regression, **MLP** – Multilayer Perceptron, **NB** – Naïve Bayes, **PLSDA** – Partial Least-Squares Discriminant Analysis, **RF** – Random Forest, **RA** – Regression Analysis, **Ridge** – Ridge Regression, **SNN** – Sequential Neural Network, **SC** – Stacking Classification, **SGB** – Stochastic Gradient Boosting, **SVM** – Support Vector Machine, **SVR** – Support Vector Regressor.

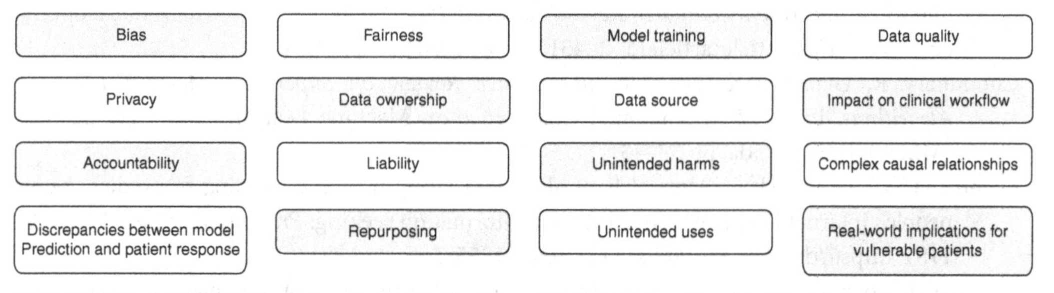

FIGURE 6.3 Ethical Considerations for the Use of ML in Reproductive Health.

likely to be incorporated into clinical decision-making, before doing so, one must take into account the impact it could have in healthcare settings by thoroughly assessing its advantages, potential misapplications as well as the issues in allocating resource's due introduction of "information noise" by it.

To ensure fairness within the ML system, various measures have been taken; these include – model constraining, the omission of sensitive variables such as race or gender from the model, the derivation outcomes that are independent of the concerned patient's identity during risk assessment, balancing of benefits and harms, among others (McCradden et al. 2020). Although the measures taken seem to be ideal, they have the potential to cause unintended harm and may create discrepancies in the detection as well as resolution of reproductive health issues; hence, designing paradigms for healthcare is a responsibility that requires meticulous deliberation and strategic planning.

6.15 CONCLUSION

ML algorithms have the potential to revolutionize the field of reproductive health by improving prediction models for outcomes, including patient mortality, hospital length of stay, or drug response to cure a sexual or reproductive disorder, as well as in enhancing decision-making processes by analyzing large datasets and detecting patterns, like patient data, medical history, diagnostic test results, and demographic information, to provide personalized recommendations for fertility treatments, prenatal care, and identifying potential complications during pregnancy, thereby assisting reproductive healthcare practitioners. Furthermore, if ML algorithms are studied and implemented on greater sample sizes and employed in larger settings in the context of reproductive medicine, they could aid in optimizing the allocation of resources and streamlining workflows in the reproductive health setting.

REFERENCES

Akinsal EC, Haznedar B, Baydilli N, et al. (2018) Artificial neural network for the prediction of chromosomal abnormalities in azoospermic males. *Urology Journal* 15:122–125. https://doi.org/10.22037/uj.v0i0.4029

Balogun JA, Egejuru NC, Idowu PA (2018) Comparative analysis of predictive models for the likelihood of infertility. *Computer Reviews Journal* 2581-6640. https://journals.indexcopernicus.com/search/article?articleId=2245130. Accessed 30 Apr 2024.

Choudhary R, Gianey HK (2017) *Comprehensive Review on Supervised Machine Learning Algorithms*. In: 2017 International Conference on Machine Learning and Data Science (MLDS). IEEE, Noida, pp 37–43.

Comulada WS, Goldbeck C, Almirol E, et al. (2021) Using machine learning to predict young people's internet health and social service information seeking. *Prevention Science* 22:1173–1184. https://doi.org/10.1007/s11121-021-01255-2

Davidson L, Boland MR (2021) Towards deep phenotyping pregnancy: a systematic review on artificial intelligence and machine learning methods to improve pregnancy outcomes. *Briefings in Bioinformatics* 22:bbaa369. https://doi.org/10.1093/bib/bbaa369

Deng X, Luo Y, Wang C (2018) *Analysis of Risk Factors for Cervical Cancer Based on Machine Learning Methods*. In: 2018 5th IEEE International Conference on Cloud Computing and Intelligence Systems (CCIS). IEEE, Nanjing, China, pp 631–635.

Denny A, Raj A, Ashok A, et al. (2019) *i-HOPE: Detection and Prediction System for Polycystic Ovary Syndrome (PCOS) Using Machine Learning Techniques*. In: TENCON 2019 - 2019 IEEE Region 10 Conference (TENCON). IEEE, Kochi, India, pp 673–678.

Dey AK, Dehingia N, Bhan N, et al. (2022) Using machine learning to understand determinants of IUD use in India: Analyses of the National Family Health Surveys (NFHS-4). *SSM - Population Health* 19:101234. https://doi.org/10.1016/j.ssmph.2022.101234

Fanton M, Nutting V, Rothman A, et al. (2022) An interpretable machine learning model for individualized gonadotrophin starting dose selection during ovarian stimulation. *Reproductive BioMedicine Online* 45:1152–1159. https://doi.org/10.1016/j.rbmo.2022.07.010

Goodson SG, White S, Stevans AM, et al. (2017) CASAnova: a multiclass support vector machine model for the classification of human sperm motility patterns†. *Biology of Reproduction* 97:698–708. https://doi.org/10.1093/biolre/iox120

Hariton E, Chi EA, Chi G, et al. (2021) A machine learning algorithm can optimize the day of trigger to improve in vitro fertilization outcomes. *Fertility and Sterility* 116:1227–1235. https://doi.org/10.1016/j.fertnstert.2021.06.018

Hoodbhoy Z, Noman M, Shafique A, et al. (2019) Use of machine learning algorithms for prediction of fetal risk using cardiotocographic data. *International Journal of Applied and Basic Medical Research* 9:226. https://doi.org/10.4103/ijabmr.IJABMR_370_18

Hossain MI, Habib MJ, Saleheen AAS, et al. (2022) Performance Evaluation of Machine Learning Algorithm for Classification of Unintended Pregnancy among Married Women in Bangladesh. *Journal of Healthcare Engineering* 2022:1–10. https://doi.org/10.1155/2022/1460908

Huang HH, Hsieh SJ, Chen MS, et al. (2023) Machine learning predictive models for evaluating risk factors affecting sperm count: predictions based on health screening indicators. *Journal of Clinical Medicine* 12:1220. https://doi.org/10.3390/jcm12031220

Islam MN, Mahmud T, Khan NI, et al. (2021) Exploring machine learning algorithms to find the best features for predicting modes of childbirth. *IEEE Access* 9:1680–1692. https://doi.org/10.1109/ACCESS.2020.3045469

Kebede SD, Mamo DN, Adem JB, et al. (2023) Machine learning modeling for identifying predictors of unmet need for family planning among married/in-union women in Ethiopia: evidence from performance monitoring and accountability (PMA) survey 2019 dataset. *PLOS Digit Health* 2:e0000345. https://doi.org/10.1371/journal.pdig.0000345

Kharb S, Joshi A (2023) Multi-omics and machine learning for the prevention and management of female reproductive health. *Frontiers in Endocrinology* 14:1081667. https://doi.org/10.3389/fendo.2023.1081667

Krenz H, Sansone A, Fujarski M, et al. (2022) Machine learning based prediction models in male reproductive health: development of a proof-of-concept model for Klinefelter Syndrome in azoospermic patients. *Andrology* 10:534–544. https://doi.org/10.1111/andr.13141

Li M, Wang H, Yang L, et al. (2020) Fast hybrid dimensionality reduction method for classification based on feature selection and grouped feature extraction. *Expert Systems with Applications* 150:113277. https://doi.org/10.1016/j.eswa.2020.113277

Li Y, Shen X, Yang C, et al. (2021) Novel electronic health records applied for prediction of pre-eclampsia: machine-learning algorithms. *Pregnancy Hypertension* 26:102–109. https://doi.org/10.1016/j.preghy.2021.10.006

Madsen A, Juul A, Aksglaede L (2023) Biochemical identification of prepubertal boys with Klinefelter syndrome by combined reproductive hormone profiling using machine learning. *Endocrine Connections* 12:e220537. https://doi.org/10.1530/EC-22-0537

Matta K, Vigneau E, Cariou V, et al. (2020) Associations between persistent organic pollutants and endometriosis: a multipollutant assessment using machine learning algorithms. *Environmental Pollution* 260:114066. https://doi.org/10.1016/j.envpol.2020.114066

Maulud D, Abdulazeez AM (2020) A review on linear regression comprehensive in machine learning. *Journal of Applied Science and Technology Trends* 1:140–147. https://doi.org/10.38094/jastt1457

McCradden MD, Joshi S, Mazwi M, Anderson JA (2020) Ethical limitations of algorithmic fairness solutions in health care machine learning. *The Lancet Digital Health* 2:e221–e223. https://doi.org/10.1016/S2589-7500(20)30065-0

Mfateneza E, Rutayisire PC, Biracyaza E, et al. (2022) Application of machine learning methods for predicting infant mortality in Rwanda: analysis of Rwanda demographic health survey 2014–15 dataset. *BMC Pregnancy Childbirth* 22:388. https://doi.org/10.1186/s12884-022-04699-8

Mohan D, Bashingwa JJH, Dane P, et al. (2019) Use of big data and machine learning methods in the monitoring and evaluation of digital health programs in India: an exploratory protocol. *JMIR Research Protocols* 8:e11456. https://doi.org/10.2196/11456

Mutlu HB, Durmaz F, Yücel N, et al. (2023) Prediction of maternal health risk with traditional machine learning methods. *NATURENGS MTU Journal of Engineering and Natural Sciences Malatya Turgut Ozal University* 4:16–23. https://doi.org/10.46572/naturengs.1293185

Naseem S, Mahmood T, Saba T, et al. (2023) DeepFert: an intelligent fertility rate prediction approach for men based on deep learning neural networks. *IEEE Access* 11:75006–75022. https://doi.org/10.1109/ACCESS.2023.3290554

Nigam A, Jaiswal P, Girkar U, et al. (2019) Migration through Machine Learning Lens – Predicting Sexual and Reproductive Health Vulnerability of Young Migrants. https://doi.org/10.48550/ARXIV.1910.02390

Ottl S, Amiriparian S, Gerczuk M, Schuller BW (2022) motilitAI: a machine learning framework for automatic prediction of human sperm motility. *iScience* 25:104644. https://doi.org/10.1016/j.isci.2022.104644

Pan L, Liu G, Mao X, et al. (2019) Development of prediction models using machine learning algorithms for girls with suspected central precocious puberty: retrospective study. *JMIR Medical Informatics* 7:e11728. https://doi.org/10.2196/11728

Raj A, Dehingia N, Singh A, et al. (2021) Machine learning analysis of non-marital sexual violence in India. *EClinicalMedicine* 39:101046. https://doi.org/10.1016/j.eclinm.2021.101046

Rúa AG, Rico N, Alonso A, et al. (2023) Ranking the effect of chronodisruption-based biomarkers in reproductive health. *Neural Computing and Applications* 35:5697–5720. https://doi.org/10.1007/s00521-022-07563-0

Sufriyana H, Husnayain A, Chen YL, et al. (2020) Comparison of multivariable logistic regression and other machine learning algorithms for prognostic prediction studies in pregnancy care: systematic review and meta-analysis. *JMIR Medical Informatics* 8:e16503. https://doi.org/10.2196/16503

Sun Q, Zou X, Yan Y, et al. (2022) Machine learning-based prediction model of preterm birth using electronic health record. *Journal of Healthcare Engineering* 2022:1–12. https://doi.org/10.1155/2022/9635526

Tesfaye B, Atique S, Azim T, Kebede MM (2019) Predicting skilled delivery service use in Ethiopia: dual application of logistic regression and machine learning algorithms. *BMC Medical Informatics and Decision Making* 19:209. https://doi.org/10.1186/s12911-019-0942-5

Thakre V (2020) PCOcare: PCOS detection and prediction using machine learning algorithms. *Bioscience Biotechnology Research Communications* 13:240–244. https://doi.org/10.21786/bbrc/13.14/56

Xu X, Yu Z, Ge Z, et al. (2022) Web-based risk prediction tool for an individual's risk of HIV and sexually transmitted infections using machine learning algorithms: development and external validation study. *Journal of Medical Internet Research* 24:e37850. https://doi.org/10.2196/37850

Yu JL, Su YF, Zhang C, et al. (2022) Tracking of menstrual cycles and prediction of the fertile window via measurements of basal body temperature and heart rate as well as machine-learning algorithms. *Reproductive Biology and Endocrinology* 20:118. https://doi.org/10.1186/s12958-022-00993-4

CHAPTER 7

Personalized Medicine in Reproductive Health

Suman Kumar Ray, Shraddha Gurha,
Suyesh Shrivastava, Vijay Pratap Singh,
Prashanth Suravajhala, and Pushpendra Singh

7.1 INTRODUCTION

Personalized medicine, a modern, technology-driven approach to individualized care, connects individuals with the best treatment based on their genetic uniqueness (Goetz and Schork 2018). This concept has emerged due to the ability to capture individual patients' genetic makeup and incorporate data outside genomes, such as epigenomics, metabolomics, and imaging, to tailor medication better. Personalized medicine emerged from the improved ability to decipher the genetic makeup of patients and match them to the optimum treatment by unraveling their genetic uniqueness and the distinct molecular makeup of their condition. As we gain access to a wider range of medicines for numerous diseases and better knowledge of molecular changes linked with the anticipated response to different treatments, personalized medicine research aims to improve our understanding and anticipate treatment efficacy based on knowledge gathered about everyone's uniqueness. Personalized medicine research aims to improve our ability to understand and anticipate treatment efficacy based on knowledge gathered about everyone's uniqueness. Personalized medicine is now extending beyond genomes to include epigenetics, proteomics, and metabolomics and furthermore, assessing environmental exposure, behaviors, and immune system may improve (Litman 2019). Even in genomics, the capacity to focus on individual cells (single-cell genomics) has revealed hitherto unseen heterogeneity (Cuomo et al. 2023). These findings corroborate observations made in patients, in which targeted medicines are beneficial for a short period of time, but individuals develop resistance because cells may contain pre-existing resistance mutations in a large cell population.

 DOI: 10.1201/9781003487548-7

The field of personalized medicine in reproductive medicine is still in its early stages of development due to its incomplete maturation. The identification and confirmation of genetic, protein, and metabolite biomarkers in the field of reproductive medicine are still in the first phases. The present approach to treating infertile patients may rely more on the subjective judgment of the attending clinician rather than the objective evidence-based facts. Medical professionals from all fields increasingly acknowledge the principle that a universal approach is not suitable for all patients. They understand that while diagnosing and treating patients, it is important to consider their individual biological characteristics (Chowers 2019).

Since the inception of the world's first baby conceived through in vitro fertilization, (IVF) more than 40 years ago, several procedures derived from IVF procedures have been developed and employed in the domain of assisted reproduction, commonly referred to as assisted reproductive technology (ART), to tackle the specific difficulties associated with male and female infertility (Steptoe and Edwards 1978). Notwithstanding these advancements, the total efficacy of ART remains significantly limited in the contemporary period, leaving ample room for enhancement in terms of cost-effectiveness, as underscored by Zhang and Yu (2020). Therefore, it is crucial to promptly tackle the issue of ART efficiency. To accomplish this goal, it is imperative to use more precise and personalized methods that are specially tailored to the unique circumstances of each couple facing infertility. Personalized medicine in the field of ART involves tailoring medicines to meet the individual requirements of each patient, rather than using a uniform approach. This tailored health model is anticipated to enhance the effectiveness, efficiency, and cost-effectiveness of treatments. Promising findings have been seen in women with reduced ovarian function, as a tailored approach to stimulating the ovaries has been shown to enhance the success of IVF, especially when employing a moderate ovarian stimulation method (Mendoza-Tesarik and Tesarik 2018). Several factors can hinder the success of ART procedures. Specific factors are associated with the quality of sperm and eggs, while different factors are related to the optimal functioning of various maternal organs. Any deviations from normal functioning in these organs can adversely affect the implantation of embryos and the subsequent growth of the fetus.

The discovery and confirmation of suitable biomarkers are required in reproductive medicine to tailor the procedure and give preventive or targeted therapy. This includes the forecasting of ovarian reserves, the result of stimulation, the quality of gametes and embryos, the receptiveness of the endometrium, pregnancy issues such early birth, and preterm labor. The current approach entails the identification of novel biomarkers by utilizing genomic, transcriptomic, proteomic, and metabolomic data stored in extensive databases, followed by validation and approval in randomized clinical trials (Palmer and Barnhart 2013).

7.2 PERSONALIZED MEDICINE AND APPROACH TO THE MANAGEMENT OF INFERTILITY

Personalized medicine depends on the concept that everyone possesses distinct molecular, metabolic, and environmental exposure characteristics. This enables customized diagnosis and therapy approaches. Although there were doubts at first regarding the practicality of customized medicine in human reproductive care, new research indicates a promising path for the future in both the short and long term (Tesarik and Mendoza-Tesarik 2022). Personalized medicine can be included into infertility treatment plans using therapeutic and biological information, even without a complete genomic characterization. The medical record consists of a detailed assessment, which encompasses a physical examination for males, a pelvic ultrasound for females, an evaluation of hormonal levels, the ages of both partners, individual and parental medical history, and any potential coexisting medical conditions (Tesarik and Mendoza-Tesarik 2022).

The comprehensive management of infertility should encompass multiple stages of decision-making, including tailored strategizing, customized preventive measures, personalized diagnostics, individualized therapies, and post-treatment monitoring. For younger individuals experiencing relatively short spells of infertility, it may be advisable to initially adopt a "wait-and-see" approach to minimize further expenses related to diagnostic testing. However, if infertility continues to be present, it is necessary to act, particularly in older individuals who have a personal and familial history of infertility, as well as lifestyle factors that may be suspected of influencing alterations in sperm production. The initial examination for the male partner should encompass a comprehensive review of medical records, a thorough physical assessment, and a study of sperm quality. It is recommended to conduct a sperm biological testing, hormonal evaluation, and seminal ultrasound evaluation in all situations. If there are identifiable warning signs for infertility or if initial assessments reveal anomalies, it is highly advisable to address them promptly. Genetic testing, testicular cytology/histology, and other sperm tests, such as the evaluation of the DNA in sperm integrity, are mostly aimed at treatment and depend on previous discoveries (Ferlin et al. 2022).

Women's reproductive dysfunction should be diagnosed with a vaginal ultrasound and simple hormonal tests. Hysteroscopy may be utilized if the individual in question has a documented history of polyps in the endometrium or prior intracavitary interventions. If both spouses' examinations reveal abnormal results, tailored preventive actions can be considered. Factors contributing to reproductive dysfunction include genetic predisposition, epigenetic abnormalities, lifestyle factors, professional exposure to gametotoxic substances, anabolics and hormones for high-level sport activities, and comorbidities like thyroid gland dysfunction or insulin resistance (Tesarik and Mendoza-Tesarik 2022). Most anomalies can be treated with suitable drugs and lifestyle adjustments. However, if no underlying cause is found, the problem becomes more difficult. Epigenetic changes that are less immediately apparent than genetic abnormalities can impair both testicular and ovarian functioning. Several previously unknown causes of disease have been associated with oxidative stress, which affects both sperm and egg cells, as well as the cells in the testes and ovaries. Dysfunction in these cells can indirectly affect the quality of gametes (Sabeti et al. 2016). Oral antioxidant treatment is gaining popularity for its effectiveness in preventing and treating certain illnesses. Nevertheless, it is crucial to acknowledge that altering the redox equilibrium from oxidative stress to reductive stress could be equally detrimental in relation to the ailment being addressed. Furthermore, reductive stress generated by excessive antioxidant consumption might be harmful to other, unrelated activities of the body.

Male fertility issues account for around half of all infertility diagnoses among couples, and many causes are unknown. Genomics and proteomics are two tools for investigating the molecular basis of male infertility. Experts believe that early detection and therapeutic development can be improved through customized medicine. Sperm analysis is a fundamental diagnostic method for determining male infertility, with sperm motility, morphology, concentration, or count being typical criteria. Human seminal plasma (HSP) contains rich molecules made by male reproductive glands, which could be useful in detecting and treating oxidative stress, which affects sperm function (Fraczek et al. 2020). Infertile guys have higher amounts of reactive oxygen species (ROS) compared to fertile men, with a prevalence ranging from 25% to 40%. Medication for male infertility appears to be effective, and some experts believe antioxidant therapy may be as well. Collaboration between researchers and physicians is crucial for developing new treatments for male infertility. Sperm DNA integrity is another personalized test for male infertility, with infertile men generally having higher amounts of sperm DNA damage. Various tests assess the overall health of sperm DNA using a sample, including the techniques used in the study like seminal chromosome structure evaluation, terminal deoxynucleotidyl transferase dUTP nick end labeling (TUNEL), and single-cell gel electrophoresis (also known as Comet assay) (Chatzimeletiou et al. 2023).

There is no reliable connection between abnormal sperm DNA and the results of pregnancy but advancements in methodology and additional studies may lead to clinical validation of these tests. The goal is to identify between fertile and infertile men soon and determine the most effective infertility therapies. Personalized medicine is making a difference in extending life by providing more effective treatment for serious illnesses, and it may also help in producing life.

7.3 WOMEN'S REPRODUCTIVE HEALTH AND PERSONALIZED MEDICINE

Table 7.1 presents major health problems and current scenarios in women's reproductive health concerning personalized medicine.

7.4 MEN'S REPRODUCTIVE HEALTH AND PERSONALIZED MEDICINE

Personalized medicine can significantly improve men's health by assisting in the identification, control, and therapy of prostate cancer, hyperplasia of the prostate, fertility problems, testosterone deficiency, and erection problems. Personalized medicine can be employed to identify individuals of the male gender who have a heightened likelihood of

TABLE 7.1 Personalized medicine in women's reproductive health.

Testing/Health Problem	Highlights
Perinatal Testing	• Non-invasive prenatal testing (NIPT) revolutionized the field by enabling non-invasive screening, as stated by Alberry et al. (2021). • NIPT, in contrast, functions primarily as a risk assessment tool that can influence care decisions, such as choosing to deliver in a specialized center, rather than serving as a true personalized medicine diagnostic connected to subsequent treatment options. • Companies are keen on broadening their operations to encompass at-home testing, wider availability, genetic screening, prenatal screening, and the identification of single gene and multiple gene danger conditions.
Pregnancy Complications	• With the increasing recognition of the elevated rate of mortality among mothers, the emphasis in the sector has shifted toward the management of complications during pregnancy. • Thermo Fisher Scientific has received approval from the FDA for its assay that assesses the risk of preeclampsia. • Currently, experts are concentrating on conducting examinations to ascertain the underlying reason of pregnancy loss. They are using AI to enhance the accuracy of ultrasound detection of fetal abnormalities. Additionally, they are exploring other approaches, such as combining RNA and AI, point-of-care methods, proteomics, and digital biomarkers, to address issues like preeclampsia and premature birth. • Only a few businesses, such MirZyme and Comanche Biopharma, can develop new treatments because it is difficult to include expectant women in clinical trials.
Endometriosis	• There is an increasing acknowledgment of the significant business potential related to endometriosis, as well as the intricate nature of the condition, especially its connections regarding impotence and the immune system. • Around 10% of females experience this illness, and it often takes around 8 years to obtain a diagnosis. There are limited non-surgical treatment options available. • Moreover, certain individuals utilize single-cell or innovative techniques for gathering, such as gadgets designed in the form of tampons. • Celmatix has recently released the most extensive genetic study on endometriosis to date. This study provides sub-typing data that has the potential to open possibilities for targeted treatment at different stages of the disease (Rahmioglu et al., 2023). • Additional pharmaceuticals now being studied include Obseva's Linzagolix, dichloroacetate, Chugai's IL-8 inhibitor, and FimmCyte's non-hormonal therapy based on antibodies. • Treatment advancements are quite limited; GnRH antagonists Orilissa and Myfembree are among the few authorized treatments, but both entail significant side effects.
Polycystic Ovary Syndrome (PCOS)	• Diagnostic activity in PCOS is relatively limited, but businesses are working to expand therapeutic choices. • Some individuals are utilizing medical devices, such as May Health's ovarian ablation technique, while other companies, like HBM Alpha therapies and Celmatix, are making progress in the development of immunoglobulin therapies.

TABLE 7.1 (Continued)

Testing/Health Problem	Highlights
Perinatal Conditions	• Currently, there is a scarcity of research being carried out on perinatal diseases such as postpartum depression also known as PPD and gestational diabetes mellitus, also known as GDM (Fischer and Morales-Suárez-Varela 2023). • Additionally, this study includes examples of digital technology used for the care of gestational diabetes mellitus, early detection of GDM, and digital therapy for postpartum depression. • The business Cedars-Sinai has discovered protein signatures in the third trimester that could be linked to mood and anxiety disorders, potentially providing new opportunities for research and treatment.
Vaginal Microbiome	• Advanced multiplex molecular diagnostics and over-the-counter (OTC) medications are enabling women to obtain a more accurate picture of their vaginal microbiota and eliminating the need for experimental treatment of illnesses with comparable signs, such as vaginosis caused by bacteria and yeast-related infections (Savicheva et al. 2023). • Several firms, like Freya Biosciences, are working toward developing immunotherapeutic treatments because of the connection between the vaginal microbiota and illnesses like a condition called end infertility, and issues with pregnancy.
Contraceptives	• Personalized contraception is an emerging trend that aims to improve the process of choosing birth control methods (Sridhar et al., 2016). • Pharmacogenomics, when used alongside testing for hormones and clinical history, has the capacity to greatly improve women's contraceptive effectiveness and reduce negative consequences.
Menopause	• Currently, there are 40 approved pharmaceuticals for menopause; however, they specifically address certain issues, and their efficacy and side effects vary. Most drugs used for steroid substitution treatments have been discovered to substantially elevate the likelihood of developing breast and ovarian cancer. • However, Astellas' newly authorized fezolinetant offers a non-hormonal option. • Moreover, there is compelling data indicating that genetics can influence the intensity of symptoms, adverse reactions, treatment efficacy, and the probability of experiencing subsequent issues. • Currently, several companies are focusing on helping during menopause by using virtual care networks or medical equipment that aid in alleviating symptoms. • However, there is still much potential for advancements in creating tailored therapies that directly target the root cause of menopause, which is the decline in ovarian function.

acquiring prostate cancer. This enables focused screening using prostate-specific antigen testing and the implementation of chemoprevention. Moreover, personalized medicine has the capability to identify individuals who do not respond to pharmacological therapy for prostatic hyperplasia, suggesting that surgery may be a more appropriate treatment choice (Mata et al. 2017). Experts in personalized medicine can advise couples on the appropriateness of testicular extraction and IVF to prevent birth defects in their children, taking into account the specific cause of their infertility. The precise impact of personalized medicine on the treatment of hypogonadism remains uncertain. However, it holds the promise of discovering certain markers linked to patients' reactions to treatment. This would enable the prescription of customized medicines (Mata et al. 2017). Personalized medicine can improve the treatment of erectile dysfunction by discovering genetic differences that control the response to drug therapies and assisting in the selection of patients for additional screening of heart conditions. Table 7.2 provides a comprehensive

TABLE 7.2 Personalized medicine in male reproductive health.

Testing/Health Problem	Highlights
Male Infertility	• Personalized medicine can improve the treatment of male-related fertility problems, which accounts for between 40% and 50% of infertility cases in couples. Men with azoospermia, a condition characterized by the absence of specimen in their ejaculation, should undergo regular clinical genetic examinations to determine the feasibility of surgical sperm retrieval (Hwang et al., 2011). • For males with severe testicular failure, the current guidelines suggest that they undergo screening for Y-chromosome microdeletions and karyotyping. • Individuals who have been diagnosed with the condition called Klinefelter syndrome (47, XXY) and have some AZFc eliminations might possess specific regions in their testes where sperm is produced. This allows for the retrieval of sperm through a procedure called intracytoplasmic sperm injection (Mata et al. 2017), which can be used to father a biological child. • Personalized medicine will facilitate the identification of sperm biomarkers that can accurately predict the fertilization efficiency of individual sperm cells (Palermo et al. 2015). • The advantages of utilizing individualized medicine for infertility therapy extend beyond intellectual aspects. They will greatly enhance couples' reproductive health and deepen their comprehension of infertility.
Hypogonadism	• The role of personalized medicine in the treatment of hypogonadism has not been established yet. • The current criteria used to diagnose hypogonadism are not accurate, and most men with symptoms are the individuals who receive androgen or (T) gel-like substances, injections, or implantation at varying levels of circulating testosterone, with varying degrees of success (Lunenfeld et al. 2015). • The diagnostic thresholds for testosterone shortage are somewhat arbitrary regarding symptoms, and it is unlikely that a definitive threshold exists for most people. • Personalized medicine has the potential to identify the most effective amounts of T (testosterone) for individual patients by analyzing metabolic indicators, hence enhancing treatment outcomes. • Personalized medicine offers the capacity to detect indicators or variations in genes that can reduce the adverse effects of T treatment and accurately predict which men are at risk of experiencing unpleasant cardiovascular events. • This technique, which represents pharmacogenomics, seeks to ensure the delivery of the suitable medication to the right patient at the ideal time. • Polymorphisms in the CYP19A1 gene, which codes for the aromatase enzyme responsible for converting testosterone to estrogen, have been demonstrated to modify levels of free testosterone (Travis et al. 2009). • Identifying this specific genetic variation in a patient with low sex hormone levels advise the clinician to prescribe an aromatase inhibitor, such as letrozole or anastrozole, as an alternative to delivering external testosterone treatment for the purpose of raising testosterone levels.

TABLE 7.2 (Continued)

Testing/Health Problem	Highlights
	• Personalized treatment may entail the analysis of repeats of CAG in the testosterone receptor gene specifically in males. This approach can aid in identifying those individuals who would derive the most value through testosterone treatment (Francomano et al. 2013). • CAG repeats were also associated with the response to testosterone treatment in a group of patients with late-onset hypogonadism who did not have surgery.
Erectile Dysfunction	• Personalized medicine shows potential for improving the medical management of erection dysfunction or ED, a common and troublesome health condition in older men. • Scientists are studying variations in gene expression to determine if they can properly predict how patients will react to inhibitors of PDE5 like sildenafil (Hatzimouratidis and Hatzichristou 2008). • Understanding these genetic differences before to pharmaceutical administration allows the doctor to select the appropriate type, dosage, and prevalence of the enzyme PDE5 inhibitor, and maybe explore alternative treatments. • The efficacy of sildenafil has already been proven in persons with pulmonary hypertension. • Patients with a genetic variant in the G-protein 3 subunit coding gene are more likely to have positive effects from sildenafil compared to those without the variation. • Personalized medicine has the potential to predict the occurrence of ED and its negative effects by analyzing genetic information. • This approach can identify genetic factors and biomarkers that contribute to the development of ED, as well as predict the likelihood of ED occurring because of radiation treatment for prostate cancer or radical prostatectomy. • This information can aid in making more informed decisions about treatment options for ED. • Personalized medicine can be particularly useful for health practitioners to determine if individuals with ED need to undergo further screening for cardiovascular disease.

summary of the primary health issues and conditions associated with men's reproductive health in the context of personalized medicine.

7.5 PERSONALIZED REPRODUCTIVE HEALTH DIAGNOSTICS AND REMEDIES

Personalized diagnostics are an extension of the procedures used to select proactive measures discussed before. An extensive inquiry has been carried out to examine the genetic and epigenetic elements linked to the occurrence of early embryo mortality (Vazquez et al. 2021). However, none of these appears to be the main cause, indicating that a combination of causes, rather than one, seems to be accountable. There is a widespread agreement that it is morally unacceptable to intentionally destroy embryos purely because they have been diagnosed with one or a few issues. This is because embryos may still possess the ability to develop normally, because of elements that are now unknown and seem to be extremely individualized. When formulating a customized treatment plan, it is essential to make two crucial decisions (Tesarik and Mendoza-Tesarik 2022). Initially, the priority of the treatment is determined, followed by the specific treatment modality selected. Both choices should be made after a thorough evaluation of the unique circumstances of each pair. The options might vary from a simple approach of observing the situation and taking minimal medication, to more complex and costly assisted reproduction therapies.

- **Selection of treatment modality:** For young couples who have experienced a relatively brief time of infertility, the most appropriate method may be to adopt a cautious and observant attitude. If there is no inherent improvement, different drugs can be utilized in conjunction with thorough clinical and laboratory monitoring to synchronize the timing of sexual intercourse with the ideal timeframe for the implantation of an embryo in the uterus. If this approach fails to achieve the intended effects, it will be necessary to utilize more sophisticated treatments, such as artificial insemination by IVF, or intracytoplasmic injection of sperm. It is crucial to acknowledge that the level of intricacy of the selected approach is directly proportional to the expense. This means that couples who opt for treatment in a private medical facility or a state-owned clinic covered by social security may incur higher costs (Tesarik and Mendoza-Tesarik 2022). Hence, it is imperative to thoroughly evaluate any prospective modifications to the selected therapeutic approach, considering the overall state of the pair under consideration.
- **Modification of the selected treatment approach:** Generally, conventional intracytoplasmic sperm injection (ICSI) is suggested for men with mild or moderate sperm abnormalities, while intracytoplasmic morphologically selected sperm injection (IMSI) is preferred for men with severe teratozoospermia or a high degree of sperm DNA fragmentation (Hazout et al. 2006). A therapy strategy, encompassing both in vivo and in vitro approaches, has been suggested, considering the severity of the illness. The treatments for this condition vary from a simple oral treatment using antioxidants to more sophisticated techniques such as ICSI and physiologic ICSI (PICSI), which involves selecting sperm based on their ability to bind hyaluronic acid, IMSI, and finally ICSI with sperm obtained through testicular biopsy – testicular sperm extraction followed by ICSI or TESE-ICSI (Tesarik and Galán-Lázaro 2017). TESE-ICSI method is recommended for individuals with severe oligoasthenoteratozoospermia or cryptozoospermia. Prior to contemplating more sophisticated and costly interventions, it is crucial to tackle the fundamental causes of male sperm abnormalities, such as hormonal imbalance or varicocele. It is recommended to administer growth hormone during ovarian stimulation for women aged 40 years or above to improve the number and quality of eggs retrieved by ovarian stimulation. It is recommended to use the same treatment for women who have a poorly developing endometrium and those who have polycystic ovarian disease. To make the best selection, it is important to carefully analyze several factors

in each case, such as the patient's age, ovarian reserve, personal and medical history, past treatment outcomes, and, if possible, information about the patient's genetic and epigenetic condition. As female age advances, it is getting more common for it to be a significant factor in human infertility. In addition, there are other more intricate indicators that can be used to identify this issue. While both physiological and premature ovarian aging lead to oxidative stress due to mitochondrial damage (Yan et al. 2022), the oocyte cytoplasm contains developmentally significant molecules that are more susceptible to oxidative stress compared to the oocyte's DNA. This is because the DNA is relatively shielded by its location within the nucleus and its association with proteins. Over 20 years ago, it was shown that transferring the nucleus from older women into donor eggs without a nucleus might maintain fertility without affecting the women's own genetic material, unlike total egg donation. To revitalize oocytes, a method was employed where a small quantity of cytoplasm from donor oocytes was injected into the patients' oocytes during IVF using ICSI (Barritt et al. 2001). Both methods have been discovered to enhance the growth of embryos and the results in patients with previous complications, including those who are advanced in age.

- **Ovarian stimulation:** Another issue to be addressed is the selection of the best ovarian stimulation protocol. We know that women respond differently to different types of hormones and their combinations (Kirshenbaum et al. 2021), and the first research into the molecular basis of these differences are emerging (Papageorgiou et al. 2021). However, for the time being, clear guidelines and protocols are still lacking, and the ovarian stimulation protocol should thus be primarily adapted based on previous attempts, in addition to repeated determinations of serum estradiol and luteinizing hormone (LH) concentrations, as previously described. Although this issue was uncovered by analyzing oocyte quality in young oocyte donors, it is even more critical in women with decreased ovarian reserve. There is an urgent need for more high-quality molecular and clinical studies to address this problem so that the best ovarian stimulation strategy can be applied as early as the first treatment attempt. Finally, in women with extremely poor ovarian response (POR) and asynchronous antral follicle growth during the follicular phase, double ovarian stimulation (DuoStim) in the follicular and luteal phases of the same cycle, followed by oocyte or embryo cryopreservation for later uterine transfer, is increasingly used. The decision-making process in cases with primary testicular disease, on the other hand, is a considerably more complicated issue with a variety of potential treatments to consider. In most cases, traditional ICSI is sufficient if spermatozoa only have minor abnormalities. In contrast, in cases of significant sperm abnormalities, IMSI or even TESE-ICSI is recommended. TESE-ICSI is especially important in extreme situations like cryptozoospermia (Tesarik and Mendoza-Tesarik 2022). IMSI, PICSI, or both in combination can help in cases with abnormally high numbers of spermatozoa with damaged DNA. Finally, in cases where oocyte activation issues have previously been reported following sperm deposition in oocyte cytoplasm, various methods of artificial oocyte activation (AOA) frequently cure the problem (Kamath et al. 2021). Different concerns are diminished ovarian reserve (DOR), issues related to advanced maternal age, poor oocyte quality, and polycystic ovary syndrome (PCOS) converge and the use of growth hormone during ovarian stimulation. The selection of a personalized ovarian stimulation protocol that adapts the doses of follicle-stimulating hormone (FSH) and LH to be administered according to the current FSH-to-LH ratio in serum at different sequential phases. Furthermore, DuoStim and meiotic spindle transfer (MST) are beneficial in individuals with DOR and other ooplasmic abnormalities (Giannelou et al. 2020).

- **Post-treatment follow-up:** An overlooked matter that remains undervalued in contemporary times is the monitoring of women after undergoing ART procedures. Recent studies refute the earlier notion that luteal phase deficit is a minor concern. These findings indicate that this problem is not limited to artificial reproductive technology cycles, but also affects natural conception cycles. Moreover, the severity of the issue is exacerbated when different ovarian stimulation regimens are employed during ART attempts. Moreover, alongside the luteal phase, the transition from the corpus luteum to the placenta as the primary source of progesterone, known as the luteo-lacental shift, seems to be disrupted in certain women following ART treatments. Administering progesterone for an extended period during pregnancy could be beneficial when serum progesterone levels start to decline after the expected completion of the luteo-placental shift. If there is no available treatment for the current primary testicular disease, it may be worth considering fertility preservation through the freezing of a sample of ejaculated sperm or surgically obtained testicular tissue. This can be done to assist in future attempts at ART. Treatment of secondary testicular failures caused by endocrine imbalances should be done concurrently with addressing the underlying cause. Failure to address hormonal imbalances can result in long-lasting disruptions in tissues and organs beyond the testis.

7.6 APPLICATION OF PERSONALIZED MEDICINE TO ASSISTED REPRODUCTION

Personalized medicine in reproductive medicine is still in its early stages due to its incomplete development. The discovery and validation of genetic, protein, and metabolite biomarkers in the discipline of reproductive healthcare is still in its early stages. The current approach to treating infertile patients is primarily guided by the subjective judgment of the attending clinician, rather than relying on the most reliable and scientifically supported data. Practitioners from all fields increasingly acknowledge the principle that a universal approach is not suitable for everyone. They understand that while diagnosing and treating patients, it is important to consider their individual molecular characteristics. This concept embodies the current approach in reproductive medicine, which involves couples undergoing multiple failed IVF rounds and subsequently repeating the same cycles. Due to the implementation of this method, we are currently facing many couples who are unable to conceive and have experienced multiple failed attempts at embryo implantation [recurrent implantation failure (RIF)] (Simón 2013).

Historical evidence suggests that the challenges of reproductive science in applying individualized therapy for promising outcomes can be resolved by transferring many embryos to boost the chances of a successful pregnancy. Simon et al. (2015) reported that this method has resulted in a significant prevalence of multiple pregnancies, with rates ranging from 30% to 40%, in IVF cycles. The consistency of infertility therapy in the field of reproductive medicine appears to be insufficient. The main differences exist in the assessment of oocytes and embryos, examination of the endometrium, analysis of sperm, and the diagnostic and therapeutic approaches used by healthcare facilities, technicians, and physicians. These variables contribute to the disparities in success rates observed between clinics and even among different clinicians within the same clinic. Reproductive medicine lacks customization due to the involvement of multiple parties, namely the mother, father, and embryos. Another concern is to the customization of several biological systems, including the egg, sperm, embryo, and uterus. It is important to acknowledge that infertility and its effective treatment in a specific marriage might be influenced by one or more systems, which can significantly impact the outcomes of the treatment (Simon et al. 2015).

Analysis of large sample groups has revealed genetic markers that are associated with natural ovulation in individuals diagnosed with PCOS. Additionally, it has been found that low-molecular-weight heparin is effective in preventing spontaneous miscarriage. Furthermore, specific variations of the FSH receptor have been linked to ovarian hyperstimulation syndrome (OHSS) and a less-than-optimal response to controlled ovarian hyperstimulation. Other biomarkers that can be utilized to adjust the dosage of gonadotropin in therapy are anti-Mullerian hormone (AMH) levels and antral follicle count (AFC). The endometrial receptivity array (ERA) is an additional biomarker that may be utilized to evaluate the gene expression in the endometrium. This biomarker offers a transcriptome signature, as stated by Jamil et al. (2016). Tailored embryo transfer is employed in individuals who have encountered recurrent failures in embryo implantation. This method entails modifying the timeframe for implanting the embryo and determining the optimal moment for transferring it.

The field of reproductive medicine necessitates the identification and validation of ideal biomarkers that individualize the entire process. These biomarkers allow for accessible preventive and targeted treatment options, such as predicting ovarian reserve, stimulation outcome, quality of reproductive cells and embryos, endometrial receptivity, ectopic pregnancy, and the possibility of preeclampsia and preterm labor.

7.7 INNOVATIONS IN PERSONALIZED MEDICINE TO IMPROVE INFERTILITY AND REPRODUCTIVE HEALTH

Various genetic, epigenomic, proteomic, metabolomic, and disease-specific biomarkers are being discovered in the field of reproductive medicine and for illnesses that impact fertility. The translation of these findings into practical applications and technologies in the field of reproductive medicine has the capacity to individualize diagnosis, customize treatment for each person, identify-specific therapies, minimize complications, and enhance patient outcomes. Instances of delayed diagnosis have been documented for a range of medical conditions, such as PCOS and endometriosis, both of which are associated with infertility (Dinsdale and Crespi 2021). Integrating genetic and biomarker findings into technological applications has the potential to expedite clinical diagnosis in individuals who are unwell, hence enabling more precise diagnosis and swifter treatment. Identifying markers linked to reduced effectiveness of fertility treatments or heightened complications, such as hyperstimulation or ectopic pregnancy, would enable clinicians to diagnose underlying conditions, choose targeted treatments with higher success rates, and more efficiently avoid treatments with negative consequences in at-risk individuals. Moreover, the discovery of genotypes associated with variations in the way individuals respond to fertility medications could result in more personalized prescriptions, perhaps leading to enhanced success rates and reduced time to achieve pregnancy and live birth. Noninvasive methods are used to evaluate male infertility and reproductive issues. Molecular biomarkers for the diagnosis and treatment of PCOS, male factor infertility, implantation failure, and first trimester miscarriage. The planned studies must exhibit the precision and safety of the device or technology across a diverse range of clinical scenarios. The objective of progressing from concept to development is to create reliable devices/technologies that can be commercially available and utilized in clinical settings.

7.8 ARTIFICIAL INTELLIGENCE AND DATA SCIENCE IN PERSONALIZED REPRODUCTIVE MEDICINE

The intersection of artificial intelligence (AI) and data science with personalized reproductive medicine marks a transformative era in healthcare, particularly in addressing the intricate challenges of reproductive health. The evolution of AI technologies and their integration into data science has opened new frontiers in personalized medicine, offering unprecedented precision in diagnosing, treating, and managing reproductive health issues (Kovac et al. 2013).

AI and machine learning (ML) algorithms have shown remarkable potential in predicting and improving the outcomes of fertility treatments, such as IVF. By analyzing vast datasets, these technologies can identify patterns and factors that influence fertility, embryo development, and pregnancy outcomes, which often elude human experts. For example, AI-driven models can analyze embryonic development stages from time-lapse images, predicting the embryos with the highest implantation potential, thus increasing the chances of successful pregnancies.

Moreover, data science applications in genomics and epigenomics are tailoring reproductive medicine to the individual's unique genetic makeup. Genetic screening and testing, facilitated by advanced computational methods, allow for the identification of genetic markers associated with fertility issues, enabling clinicians to offer personalized treatment plans. This approach not only enhances the efficacy of treatments but also minimizes the risks and emotional strain associated with fertility procedures.

In personalized reproductive health, AI and data science also play crucial roles in enhancing the understanding of complex conditions like PCOS and endometriosis. These technologies enable the aggregation and analysis of diverse data types, including clinical, genomic, and lifestyle information, offering insights into the etiology and progression of these conditions. Consequently, this leads to more effective, individualized management strategies that can improve reproductive outcomes and overall quality of life for patients.

7.9 LIMITATIONS AND CHALLENGES IN PERSONALIZED REPRODUCTIVE HEALTH

Despite the promising advancements, the integration of AI and data science in personalized reproductive medicine faces significant challenges. Privacy and ethical concerns top the list, as the handling of sensitive genetic and health data requires stringent safeguards to prevent misuse and ensure confidentiality. Furthermore, the complexity and variability of reproductive health issues necessitate models that can adapt to a wide range of scenarios, which is challenging given the current limitations in AI and ML algorithms.

Another critical challenge is the digital divide and access to technology. The benefits of personalized medicine, driven by AI and data science, are yet to be universally accessible, with disparities in healthcare infrastructure and technology adoption across different regions. This gap limits the widespread application of these innovations, leaving a significant portion of the population without access to these advanced treatments.

Lastly, the interpretation and integration of AI and data science findings into clinical practice present hurdles. The complexity of data-driven insights requires clinicians to have a deep understanding of these technologies, which is not always the case. Bridging this knowledge gap is essential for the effective translation of AI and data science advancements into improved reproductive health outcomes.

7.10 LONG-TERM CONSEQUENCES

The long-term consequences of integrating AI and data science into personalized reproductive medicine are profound. As these technologies evolve, they promise to enhance our understanding of reproductive health, leading to earlier diagnoses, more effective treatments, and improved prognostic capabilities. This shift toward more personalized and precise medicine could significantly increase the success rates of fertility treatments, reduce the incidence of genetic diseases, and improve the overall well-being of individuals and families.

However, the long-term success of these technologies hinges on addressing the current limitations and challenges. Ensuring privacy, expanding access, and fostering interdisciplinary collaboration between technologists and clinicians will be critical. Moreover, as AI and data science continue to reshape the landscape of reproductive medicine, ongoing ethical considerations and societal impacts must be carefully managed.

7.11 CONCLUSION AND FUTURE PERSPECTIVES

Personalized medicine, a field of medicine that aims to tailor treatment to individual patients beyond the present standard of care, is seeing increasing popularity. AI and data science, which have lately achieved maturity, will have a growing significance in expanding the scope of personalized treatment. In the future, personalized medicine and AI will have significant roles in the IVF clinic. They will not only improve outcomes but also reduce pregnancy complications and provide couples with greater control over their reproductive process. The favorable rates for aided human reproduction are considerably inferior when compared to those of other animals. This can be partially attributed to the comparatively reduced reproductive efficiency inherent in the human species when compared to animals. Another variable to consider is the increasing mean age of women. However, these two elements should not be predominantly ascribed. Increasing data suggests that each infertile couple has unique needs that must be considered while selecting the most suitable diagnostic, clinical, and laboratory treatments, as well as post-treatment monitoring. The individualized approach is becoming increasingly essential as the situation grows more intricate. While additional thorough molecular and clinical investigation is necessary to successfully apply personalized medicine principles in human-assisted reproduction, we can presently employ information from the patient's medical history, previous attempts, and available genetic and epigenetic data to tailor the treatment for each couple's particular condition in the most advantageous way.

Implementing a tailored medicine strategy for men's health will enhance the identification and management of diverse medical conditions. It is essential for fundamental scientists, doctors, engineers, pathologists, statisticians, and patients to work together to create biomarkers that are individual to each patient. This collaboration is necessary to increase the range of treatment options that clinicians have access to. To completely implement personalized medicine, it is imperative that we engage in translational research, promote greater male participation in clinical and genetic research, and ensure that marginalized regions have access to personalized medicine technologies, thereby ensuring equal accessibility for all men. We have a positive outlook on the potential growth and transformation of this emerging industry in the next years, which will have significant advantages for men and their families worldwide.

REFERENCES

Alberry MS, Aziz E, Ahmed SR, Abdel-fattah S (2021) Non invasive prenatal testing (NIPT) for common aneuploidies and beyond. *European Journal of Obstetrics & Gynecology and Reproductive Biology* 258:424–429. https://doi.org/10.1016/j.ejogrb.2021.01.008

Barritt JA, Willadsen S, Brenner C, Cohen J (2001) Cytoplasmic transfer in assisted reproduction. *Human Reproduction Update* 7:428–435. https://doi.org/10.1093/humupd/7.4.428

Chatzimeletiou K, Fleva A, Nikolopoulos TT, et al. (2023) Evaluation of sperm DNA fragmentation using two different methods: TUNEL via fluorescence microscopy, and flow cytometry. *Medicina* 59:1313. https://doi.org/10.3390/medicina59071313

Chowers Y (2019) One size does not fit all: the case for translational medicine. *Rambam Maimonides Medical Journal* 10:e0011. https://doi.org/10.5041/RMMJ.10364

Cuomo ASE, Nathan A, Raychaudhuri S, et al. (2023) Single-cell genomics meets human genetics. *Nature Reviews Genetics* 24:535–549. https://doi.org/10.1038/s41576-023-00599-5

Dinsdale NL, Crespi BJ (2021) Endometriosis and polycystic ovary syndrome are diametric disorders. *Evolutionary Applications* 14:1693–1715. https://doi.org/10.1111/eva.13244

Ferlin A, Calogero AE, Krausz C, et al. (2022) Management of male factor infertility: position statement from the Italian Society of Andrology and Sexual Medicine (SIAMS): Endorsing Organization: Italian Society of Embryology, Reproduction, and Research (SIERR). *Journal of Endocrinological Investigation* 45:1085–1113. https://doi.org/10.1007/s40618-022-01741-6

Fischer S, Morales-Suárez-Varela M (2023) The bidirectional relationship between gestational diabetes and depression in pregnant women: a systematic search and review. *Healthcare* 11:404. https://doi.org/10.3390/healthcare11030404

Fraczek M, Wojnar L, Kamieniczna M, et al. (2020) Seminal plasma analysis of oxidative stress in different genitourinary topographical regions involved in reproductive tract disorders associated with genital heat stress. *International Journal of Molecular Sciences* 21:6427. https://doi.org/10.3390/ijms21176427

Francomano D, Greco EA, Lenzi A, Aversa A (2013) CAG repeat testing of androgen receptor polymorphism: is this necessary for the best clinical management of hypogonadism? *The Journal of Sexual Medicine* 10:2373–2381. https://doi.org/10.1111/jsm.12268

Giannelou P, Simopoulou M, Grigoriadis S, et al. (2020) The conundrum of poor ovarian response: from diagnosis to treatment. *Diagnostics* 10:687. https://doi.org/10.3390/diagnostics10090687

Goetz LH, Schork NJ (2018) Personalized medicine: motivation, challenges, and progress. *Fertility and Sterility* 109:952–963. https://doi.org/10.1016/j.fertnstert.2018.05.006

Hatzimouratidis K, Hatzichristou DG (2008) Looking to the future for erectile dysfunction therapies. *Drugs* 68:231–250. https://doi.org/10.2165/00003495-200868020-00006

Hazout A, Dumont-Hassan M, Junca AM, et al. (2006) High-magnification ICSI overcomes paternal effect resistant to conventional ICSI. *Reproductive BioMedicine Online* 12:19–25. https://doi.org/10.1016/S1472-6483(10)60975-3

Hwang K, Lipshultz LI, Lamb DJ (2011) Use of diagnostic testing to detect infertility. Current Urology Reports 12:68–76. https://doi.org/10.1007/s11934-010-0154-0

Jamil Z, Fatima SS, Ahmed K, Malik R (2016) Anti-mullerian hormone: above and beyond conventional ovarian reserve markers. *Disease Markers* 2016:1–9. https://doi.org/10.1155/2016/5246217

Kamath MS, Vogiatzi P, Sunkara SK, Woodward B (2021) Oocyte activation for women following intracytoplasmic sperm injection (ICSI). *Cochrane Database of Systematic Reviews* 2021. https://doi.org/10.1002/14651858.CD014040

Kirshenbaum M, Gil O, Haas J, et al. (2021) Recombinant follicular stimulating hormone plus recombinant luteinizing hormone versus human menopausal gonadotropins – does the source of LH bioactivity affect ovarian stimulation outcome? *Reproductive Biology and Endocrinology* 19:182. https://doi.org/10.1186/s12958-021-00853-7

Kovac JR, Pastuszak AW, Lamb DJ (2013) The use of genomics, proteomics, and metabolomics in identifying biomarkers of male infertility. *Fertility and Sterility* 99:998–1007. https://doi.org/10.1016/j.fertnstert.2013.01.111

Litman T (2019) Personalized medicine – concepts, technologies, and applications in inflammatory skin diseases. *APMIS* 127:386–424. https://doi.org/10.1111/apm.12934

Lunenfeld B, Mskhalaya G, Zitzmann M, et al. (2015) Recommendations on the diagnosis, treatment and monitoring of hypogonadism in men. *The Aging Male* 18:5–15. https://doi.org/10.3109/13685538.2015.1004049

Mata DA, Katchi FM, Ramasamy R (2017) Precision medicine and men's health. *American Journal of Men's Health* 11:1124–1129. https://doi.org/10.1177/1557988315595693

Mendoza-Tesarik R, Tesarik J (2018) Usefulness of individualized FSH, LH and GH dosing in ovarian stimulation of women with low ovarian reserve. *Human Reproduction* 33:981–982. https://doi.org/10.1093/humrep/dey063

Palermo G, Neri Q, Rosenwaks Z (2015) To ICSI or not to ICSI. *Seminars in Reproductive Medicine* 33:092–102. https://doi.org/10.1055/s-0035-1546825

Palmer SS, Barnhart KT (2013) Biomarkers in reproductive medicine: the promise, and can it be fulfilled? *Fertility and Sterility* 99:954–962. https://doi.org/10.1016/j.fertnstert.2012.11.019

Papageorgiou K, Mastora E, Zikopoulos A, et al. (2021) Interplay between mTOR and Hippo Signaling in the ovary: clinical choice guidance between different gonadotropin preparations for better IVF. *Frontiers in Endocrinology* 12:702446. https://doi.org/10.3389/fendo.2021.702446

Rahmioglu N, Mortlock S, Ghiasi M, et al. (2023) The genetic basis of endometriosis and comorbidity with other pain and inflammatory conditions. *Nature Genetics* 55:423–436. https://doi.org/10.1038/s41588-023-01323-z

Sabeti P, Pourmasumi S, Rahiminia T, et al. (2016) Etiologies of sperm oxidative stress. *International Journal of Reproductive BioMedicine* 14:231–240. https://doi.org/10.29252/ijrm.14.4.231

Savicheva AM, Krysanova AA, Budilovskaya OV, et al. (2023) Vaginal microbiota molecular profiling in women with bacterial vaginosis: a novel diagnostic tool. *International Journal of Molecular Sciences* 24:15880. https://doi.org/10.3390/ijms242115880

Simón C (2013) Personalized assisted reproductive technology. *Fertility and Sterility* 100:922–923. https://doi.org/10.1016/j.fertnstert.2013.08.011

Simon C, Sakkas D, Gardner DK, Critchley HOD (2015) Biomarkers in reproductive medicine: the quest for new answers. *Human Reproduction Update* 21:695–697. https://doi.org/10.1093/humupd/dmv043

Sridhar A, Rodriguez J, Roque K, Abutouk M (2016) Personalized contraceptive counseling: helping women make the right choice. Open Access Journal of Cardiology 7:89–96. https://doi.org/10.2147/OAJC.S81546

Steptoe PC, Edwards RG (1978) Birth after the reimplantation of a human embryo. *The Lancet* 312:366. https://doi.org/10.1016/S0140-6736(78)92957-4

Tesarik J, Galán-Lázaro M (2017) Clinical scenarios of unexplained sperm DNA fragmentation and their management. *Translational Andrology and Urology* 6:S566–S569. https://doi.org/10.21037/tau.2017.03.70

Tesarik J, Mendoza-Tesarik R (2022) Patient-tailored reproductive health care. Frontiers in Reproductive Health 4:917159. https://doi.org/10.3389/frph.2022.917159

Travis RC, Schumacher F, Hirschhorn JN, et al. (2009) CYP19A1 genetic variation in relation to prostate cancer risk and circulating sex hormone concentrations in men from the Breast and Prostate Cancer Cohort Consortium. *Cancer Epidemiology, Biomarkers & Prevention* 18:2734–2744. https://doi.org/10.1158/1055-9965.EPI-09-0496

Vazquez MJ, Daza-Dueñas S, Tena-Sempere M (2021) Emerging roles of epigenetics in the control of reproductive function: focus on central neuroendocrine mechanisms. *Journal of the Endocrine Society* 5:bvab152. https://doi.org/10.1210/jendso/bvab152

Yan F, Zhao Q, Li Y, et al. (2022) The role of oxidative stress in ovarian aging: a review. *Journal of Ovarian Research* 15:100. https://doi.org/10.1186/s13048-022-01032-x

Zhang PY, Yu Y (2020) Precise personalized medicine in gynecology cancer and infertility. *Frontiers in Cell and Developmental Biology* 7:382. https://doi.org/10.3389/fcell.2019.00382

CHAPTER 8

Ethics and Privacy Considerations

Puja Singh and Satish Singh

8.1 INTRODUCTION

8.1.1 Confidentiality and Privacy

8.1.1.1 Confidentiality

Confidentiality is the ethical and legal responsibility to protect information shared by an individual with a professional or an organization. This information should not be disclosed to unauthorized persons without the consent of the person who gave it.

8.1.1.2 Privacy

Privacy is the right of individuals to keep their personal information protected from unwarranted intrusion. It encompasses the idea that individuals have control over who can access their details and under what circumstances.

- **Patient Confidentiality**

Patient confidentiality is one of the pillars of ethical healthcare practices, especially in reproductive health. Due to the sensitive and personal nature of reproductive health information, a heightened commitment is required to preserve patient privacy. This commitment goes beyond protecting medical records and includes wider aspects such as reproductive decision-making, family planning, and potential stigmatization associated with certain conditions.

Patient confidentiality in reproductive health covers a variety of sensitive information. It covers from fertility problems and pregnancy status to sexually transmitted diseases and genetic traits. According to Brown et al (2019), confidentiality is crucial to building

DOI: 10.1201/9781003487548-8

trust between healthcare providers and patients. It encourages open communication and facilitates informed decision-making.

The reproductive health of adolescents adds complexity to patient confidentiality. Health professionals must strike a balance between respect for the privacy rights of young children and the appropriate involvement of parents or guardians. Saenz et al (2017) stated that the legal policies and ethical guidelines should include specific circumstances in which privacy may be breached, such as when a teenager or others are harmed.

An additional new dimension adds to the complexity of patient privacy in cases involving assisted reproductive technologies (ART), or surrogacy. The protection of information of donors and surrogate mothers is critical for their reputation. Complex legal systems related to reproductive healthcare make them even more challenging.

- **Protection of Sensitive Health Information**

Protection of medical records is an important aspect of patient privacy in reproductive health. Medical records hold vital reproductive health information such as birth logs, prenatal care, abortion services, and genetic test findings. Enhanced approaches should be used to organize both paper and electronic records. To ensure the privacy and security of these records, security assessments, access controls, and encryption procedures must be implemented.

Health professionals are involved in discussions on family planning, fertility problems, and intimate aspects of sexual health. Creating an environment in which patients feel comfortable sharing their personal information is essential. It fosters open communication and ensures that individuals will most likely seek the care they need. Thus, establishing a safe and confidential space for reproductive health discussions is of utmost importance.

- **Access Controls to Health Records**

Access controls are essential for protecting sensitive medical information in health records. They function as an impediment, permitting access to medical records exclusively to authorized personnel. To validate the identity of individuals seeking access, strong and secure authentication techniques such as usernames, passwords, biometrics, or smart cards be implemented.

In addition, the establishment of role-based access controls is a common practice in restricting the information accessible to healthcare professionals according to their specific roles and responsibilities. The use of encryption and audit trails adds an extra layer of protection to health records. Careful tracking and recording of who accessed the information, and for what purpose helps to control and prevent data breach. These measures play a vital role in upholding patient confidentiality, complying with privacy regulations, and effectively managing healthcare data with integrity.

8.2 INFORMED CONSENT

A patient's informed consent is a purposeful and unique agreement that is accompanied by detailed and comprehensive information about the benefits, drawbacks, advantages, and alternatives associated with a medical procedure or the use of personal data for purposes such as education and research, among other things. Informed consent is based on the principle of respecting individual autonomy and privacy. It is inextricably related to the legal system.

Informed consent includes the following essential attributes and practices:

(a) **Timely and Prior Information:** Patients must be given thorough information before any medical procedure. They should be allowed to ponder calmly, seek clarification, and make educated judgments without feeling rushed.

(b) **Clear and Understandable Information:** To facilitate informed decision-making details of the treatment or procedure should be delivered in clear and understandable language avoiding medical jargon.

(c) **Information Personalization:** Recognize the uniqueness of each patient. Information should be customized to reflect the individual's health knowledge, culture, and personal preferences.

(d) **Use of Multimedia and Visual Aids:** It is said that a picture says a thousand words. Similarly, visual aids such as images, videos, or booklets may be used to increase audience interaction and engagement. They can simplify complex reproductive health-related medical conditions or procedures.

(e) **Provide Ample Time:** Allow plenty of time for patients to understand the content and seek clarification at their own pace. Haste hampers the patient's understanding and limits the ability to make educated choices.

(f) **Addressing Questions and Concerns:** Create an atmosphere that allows patients to freely raise any questions or concerns they may have. Healthcare practitioners must be collaborative and composed.

(g) **Discuss Risks and Benefits:** A balanced approach offering potential risks and benefits should be adopted. This will allow patients to make decisions based on sound understanding.

(h) **Alternative Options:** Provide options wherever feasible. It allows people to actively engage in decision-making and align their choices with beliefs and preferences.

(i) **Repetition and Summarization:** Identify key points and summarize important information. Repetition reinforces important information, helping patients retain and understand information better.

(j) **Documentation of Understanding:** Let patients explain the facts in their own words. Simple question-and-answer forms can be employed.

(k) **Incorporate Collaborative Decision-Making:** Patients should be encouraged to actively engage in the decision-making process. This will help them incorporate their priorities, values, and preferences.

(l) **Capacity Assessment:** Healthcare practitioners must assess the patient's decision-making ability. If needed alternative legal decision-making mechanisms should be adopted.

(m) **Patient's Right to Refuse:** The patient should feel confident in declining any treatment or procedure without fear of any repercussions.

(n) **Consent Forms:** Use written permission forms for consent before any clinical procedure. Such a form should cover a detailed summary of potential dangers, advantages, and alternatives.

(o) **Cultural Competence:** Acknowledge and honor cultural variations that could affect a patient's choice of treatment. Certain cultures place a high value on group decision-making, including agreement from family members.

(p) **Pediatric and Adolescent Consent:** Get parents' or legal guardians' informed consent before treating youngsters. When dealing with older kids, evaluate the adolescent's capacity for decision-making. Involve them in the consent procedure while taking ethical and legal considerations into account.

(q) **Emergency Conditions:** Healthcare professionals should adhere to established processes. If possible, seek verbal consent in emergencies where getting written consent might not be possible. It is crucial to record the situation, the need for the intervention, and any attempts at contact with the patient or proxy.

(r) **Secondary Uses and Future Research:** Seek proper consent from the participants after explaining clearly the prospective secondary use or future research.

(s) **Ethical Review Board Approval:** Appropriate approval from an institutional or ethical review board (IRB) should be taken to ensure that the research procedure meets prescribed ethical criteria.

(t) **Data Security and Governance:** Governance frameworks and security procedures should be in place to protect the data. This covers access controls, encryption, and following rules and laws about data protection.

- **Ensuring Fully Informed Decision-Making**

Fully informed decision-making in healthcare is pivotal for fostering patient autonomy. It promotes trust and enhances the overall quality of healthcare delivery. With a complete understanding of their medical conditions, available treatments, and potential risks, individuals feel engaged in decision-making about their health. This not only promotes autonomy but also lets them align choices with personal values and preferences (Brezina and Zhao 2012).

Muhunthan and Arulkumaran (2014) concluded that fully informed decision-making cultivates a stronger relationship built on collaboration and transparency. This creates a positive healthcare environment that aids in better treatment adherence, leading to increased patient satisfaction and ultimately, improved health outcomes.

Healthcare staff must use unambiguous language to explain complex medical jargon to patients. Healthcare providers must address any concerns and apprehensions patients have to enable them to make choices that align with their best interests.

- **Consent for Treatment and Procedures**

Obtaining consent for medical treatment and procedures is a vital ethical and legal principle in the healthcare industry. It emphasizes the utmost regard for an individual's autonomy and their ability to make informed choices. This provides patients with comprehensive information on the specifics, objectives, potential risks, and alternative options. This technique fosters transparency, confidence, and a collaborative relationship between patients and healthcare providers, facilitating a shared decision-making process.

Furthermore, consent also serves as a legal safeguard for healthcare providers against any legal ramifications while ensuring that all medical procedures are conducted in an ethical manner and per the patient's wishes. Ultimately, informed consent not only respects the dignity and rights of individuals but also improves the overall standard and trustworthiness of healthcare systems.

- **Consent for Data Sharing and Research**

A substantial volume of confidential and classified data is collected as part of reproductive health. There is tremendous potential this data has in data sharing and research. It can lead to significant technical advances, rapid healthcare delivery, and mercurial response to medical calamities. However, using such data without due consent from patients/parents/guardians will not only be a breach of patients' autonomy but also can have legal implications.

Reproductive health professionals must communicate the purpose and potential impacts of collecting and sharing reproductive health information. Patients should be well informed about how their data may be used for medical research or educational purposes. Patients ought to have been duly apprised of their right to give informed consent or refuse involvement in the aforementioned study.

Based on comprehensive comprehension, patients, parents, or custodians will be more inclined to make informed decisions as on comfort and inclinations. Thus, a system should be implemented that would allow patients to specify whether they want data to be shared with other healthcare professionals or for research purposes.

8.3 DATA SECURITY AND STORAGE

Prioritizing the privacy and security of reproductive health information is crucial to maintain trust and adhere to privacy regulations. Similarly, implementing strong data security measures, such as encryption, access controls, and frequent audits, is essential in preventing any potential unauthorized access or breaches. Additionally, all storage solutions, whether digital or physical, should comply with industry standards and legal requirements to protect them against potential data loss or theft (Premkumar 2018).

Adhering to ethical standards and maintaining confidentiality as a top priority not only secures sensitive reproductive health data but also maintains the integrity of the healthcare system. This helps build trust and security between patients and healthcare providers. Thus, establishing suitable standards and protocols for data storage and administration guarantees the responsible and ethical handling of reproductive health information (Schweikart 2018).

- **Secure Handling of Reproductive Health Data**

Stringent security measures are needed to protect health information from unauthorized access or breaches. These measures include adhering to data protection laws and regulations as well as access controls, secure databases, encryption, and regular cybersecurity audits.

Some important measures for handling data securely are:

(a) **Regulatory Compliance:** It is essential to acknowledge and adhere to relevant laws and guidelines regarding data protection, such as the renowned General Data Protection Regulation (GDPR) in the European Union or the Health Insurance Portability and Accountability Act (HIPAA) in the United States.

(b) **Data Encryption:** Appropriate encryption technologies should be used to safeguard data while it's in transit or at rest.

(c) **Access Controls:** Rigorous access controls should be in place to safeguard sensitive information regarding reproductive health. Appropriate procedures should be established and enforced to restrict access to only authorized personnel.

(d) **Authentication Mechanism:** Robust authentication techniques, such as two-factor authentication, should be employed to validate users' identity before granting access.

(e) **Anonymization and Pseudonymization:** Pseudonymization or anonymization methods should be appropriately used to modify or eradicate personally identifiable information (PII).

(f) **Regular Audit Trails:** Keep through audit record logs of who, when, and what modifications were done. Regular reviews of these logs help detect any unauthorized or suspicious activities.

(g) **Secure Storage Solutions:** Opt for secure storage solutions, such as encrypted databases and cloud storage with comprehensive security features. Verify that storage provider's adherence to industry standards and regulations.

(h) **Data Minimization:** Only gather and retain the bare minimum of information required for reproductive health. Refrain from retaining superfluous data to reduce the chance of unwanted access.

(i) **Regular Backups:** Device malfunctions, cyberattacks, or other unanticipated circumstances can cause data loss. Thus, data should be appropriately backed regularly. The restoration process should be periodically tested to ensure its effectiveness.

(j) **Secure Communication Protocols:** To ensure the safe transit of data across systems and organizations secure communication protocols, such as Hypertext Transfer Protocol Secure (HTTPS) should be employed.

(k) **Physical Security:** Security measures such as access controls, and surveillance should be in place against theft or unauthorized access to hardware storing hardware data.

(l) **Employee Training:** Thorough training on privacy regulations, data security best practices, and the value of protecting sensitive data should be imperative for personnel handling sensitive reproductive health data.

(m) **Incident Response Plan:** An incident response strategy should be created and updated regularly to handle possible security events and data breaches. It should cover communication channels and specify protocols in case of security incidents.

(n) **Regular Software Updates and Patching:** All systems and software should be updated with the most recent security updates to fix any vulnerabilities.

(o) **Secure Data Disposal:** Protocols should be implemented to securely destroy data that is no longer needed.

(p) **Vendor Security:** Third-party vendors who handle data related to reproductive health must comply with rigorous security protocols. The security of their systems should be regularly evaluated.

(q) **Privacy Impact Assessments:** Privacy impact assessments should be conducted to identify and mitigate any privacy concerns that may arise from the collection, retention, and manipulation of reproductive health data.

- **Storage and Retrieval Safeguards**

Electronic health records (EHR) systems must have advanced encryption mechanisms to protect sensitive information. Rigorous authentication protocols and strict access controls should be in place to avoid unauthorized access and to maintain patient confidentiality, thus effectively minimizing the threat of potential data breaches.

Many governments are trying to put acts and laws to enforce sensitive data storage and retrieval safeguards. In India, the Digital Information Security in Healthcare Act (DISHA) is one such act aimed at creating a national digital health authority and health information exchange. The proposed regulation aims to improve privacy, confidentiality, security, and standardization of electronic health information.

Quick and accurate access to crucial information is vital for healthcare professionals. Retrieval safeguards will be pivotal and challenging to seamlessly exchange information between various healthcare entities, including hospitals, clinics, and laboratories. Therefore, establishing standards and protocols for data exchange is imperative to ensure efficient retrieval without compromising the security and reliability of the information.

Viable strategies to ensure strict retrieval safeguard protocols for reproductive healthcare information include training, role-based access, authentication and authorization, the need-to-know principle, secure patient portals, temporary access for specific tasks, health record encryption, and the use of comprehensive statistical methods to detect data breaches and malicious activities.

Data loss due to unforeseen situations can be avoided by adopting thorough backup and disaster recovery strategies. Frequent backup processes and contingency plans should be implemented. This will not only ensure the continuity of healthcare services but also minimize any disruptions caused by system breakdowns or natural catastrophes.

In the end, honoring the autonomy and respect of every individual is of foremost importance. A thorough and diverse strategy is needed to efficiently manage reproductive healthcare data. Healthcare institutions need to strike a harmonious balance between accessibility and confidentiality. They can achieve this goal by implementing cutting-edge encryption, strict access measures, compatible EHR systems, legal adherence, and contingency protocols. This will not only boost their credibility with patients but also promote the evolution of reproductive healthcare.

8.4 GENETIC INFORMATION AND COUNSELING

Genetic information in reproductive healthcare plays a very crucial role. It can provide vital information on potential hereditary risks and predispositions. Through genetic counseling by skilled professionals, the crucial link between science and informed decision-making can be established. Such specialized forms of counseling can help individuals and couples comprehend the implications of their genetic information and the likelihood of passing on genetic conditions to their children (Shaw 2004).

Genetic counselors enable people to make well-informed decisions about their reproductive choices by thoroughly delving into the family background, interpreting genetic test outcomes, and addressing potential risks. Potential alternatives suggested can involve prenatal testing, ART, and adoption. The amalgamation of genetic data and counseling facilitates the creation of personalized treatment plans and enables well-informed family planning decisions. Ultimately, this promotes the overall welfare of individuals and their future offspring.

- **Genetic Counseling Practices**

Some key genetic counseling techniques are as follows:

(a) **Comprehensive Assessment:** The foundation of genetic counseling lies in the thorough evaluation of the individual or couple's medical history, family history, and relevant genetic testing results.

(b) **Clear Communication:** Genetic counselors must convey complex genetic information clearly and avoid jargon usage. Clients should have a solid grasp of the implications and options available.

(c) **Informed Consent:** Genetic counselors must ensure that individuals fully understand the purpose, benefits, and limitations of genetic testing.

(d) **Emotional Support:** Genetic counselors should explain the potential emotional and psychological implications of genetic findings.

(e) **Privacy and Confidentiality:** Professionals must adhere to legal and ethical standards for safeguarding the sensitive nature of genetic information. Genetic information should be disclosed only with the explicit consent of the individual or couple.

(f) **Cultural Sensitivity:** Understanding cultural nuances helps in tailoring counseling approaches and recommendations. This helps in aligning genetic counseling with the unique beliefs and preferences of clients.

(g) **Continuing Education:** Genetic counselors must engage in ongoing education to stay abreast of the latest developments in the field.

(h) **Collaboration with the Healthcare Team:** Genetic counselors should work collaboratively with other healthcare professionals, including obstetricians, geneticists, etc. This will help them gain a holistic understanding of the individual's health and facilitate comprehensive care.

In addition, Genetic counselors should provide post-counseling support. Everyone should have equitable access to genetic counseling in reproductive healthcare, regardless of cultural, economic, or geographical barriers. Therefore, it is critical to reduce these impediments.

- **Implications and Consent for Genetic Testing**

The implications of genetic testing in reproductive healthcare are significant for both individuals and couples. Some of the extensive ramifications of genetic testing outcomes are the identification of genetic disorders, risk assessments for progeny, prenatal diagnosis, personalized treatment strategies, emotional and psychological impact, effects on family dynamics and relationships, financial implications, lifestyle choices, and legal and insurance considerations (Christianson et al 2022).

Considering these implications, informed consent necessitates candid and transparent communication. Such communication should cover all potential consequences of genetic results. Individuals must understand the range of possible outcomes, including discovering genetic conditions, the influence on future generations, and the potential emotional impact. Any informed consent should be freely given and not coerced. Individuals should be free to align their choices with their values and preferences. As the realm of genetic testing is continuously evolving, healthcare professionals must uphold ethical standards to safeguard the rights and welfare of individuals.

- **Ethical Handling of Genetic Data**

As the landscape of reproductive healthcare is redefined with the advances in genetic technologies, ethical considerations must guide every stage of data collection, storage, and utilization. Informed consent becomes imperative for any genetic testing. Similarly, confidentiality and privacy protection are equally crucial. Healthcare providers and researchers are obligated to take robust security measures to prevent unauthorized access or disclosure of sensitive genetic information.

Individuals can make decisions aligned with their values and preferences when potential risks and limitations of genetic testing are communicated transparently. There is an imperative need for clear guidelines and mechanisms to protect participants' genetic data for responsible and equitable use in research. Maintaining the right balance between scientific progress and ethical principles is essential to fostering trust in the field of reproductive health and genetic data management. Thus, ethical handling of genetic data in reproductive health is of paramount importance to safeguard individual privacy, autonomy, and dignity.

8.5 REPRODUCTIVE RIGHTS AND AUTONOMY

Freedom to make educated decisions about one's own health and reproductive options is a key component of reproductive rights. To make informed decisions, access to accurate and exhaustive information regarding the risks, benefits, and alternatives is vital. This idea, built upon principles of human rights, is universally accepted internationally.

All individuals have the autonomy to make reproductive health decisions, free from any coercion, prejudice, or violence (Davis 2022). To achieve these objectives, individuals ought to possess convenient access to family planning services, and techniques such as family planning decisions, fertility counseling, and contraception options. Infertility treatment patients should have comprehensive knowledge of procedures, associated success rates, and ethical considerations associated. Access to ART and freedom of their proper application is a fundamental principle of reproductive autonomy. Reproductive autonomy also encompasses legal and secure abortion services.

Programs to spread information and education about sexually transmitted illnesses, consent, contraception, and the anatomy of reproductive health should be run. Legal frameworks can guarantee access to information and resources as needed and that individual autonomy be honored. These will be the bedrock to safeguard people's ability to make educated decisions about their reproductive health.

- **Ethical Considerations in Assisted Reproductive Technologies**

Ethical considerations in ART encompass a complex interplay of values, rights, and societal norms. These must guide the development and application of these innovative medical interventions. The principle of autonomy sits right in the middle of these considerations.

Many ethical concerns are voiced about ART techniques such as in vitro fertilization (IVF), surrogacy, and genetic screening. To prevent disparities, it is vital to have equitable access to these technologies, regardless of socioeconomic status. Additional ethical challenges are posed by issues surrounding the moral status of embryos, the potential for multiple pregnancies, and the responsible use of genetic information. The physical and mental well-being of participants is essential. Thus potential risks should be transparently disclosed to them (Schenker and Eisenberg 1997). The ever-evolving landscape of ART demands continuous ethical scrutiny to maintain a delicate balance between promoting reproductive autonomy and safeguarding the well-being of individuals, couples, and the potential offspring involved.

8.6 EQUITY AND ACCESS TO REPRODUCTIVE HEALTH

Reproductive health is extremely crucial for gender equality, personal well-being, and global development. Nonetheless, disparities in access are an abhorrent reality within this critical field. Reproductive health equality and access require a paradigm shift that addresses a multitude of obstacles and closes gaps that ensnare innumerable individuals (Cook et al 2003).

The impact of the difficulties faced in equity and access to reproductive health varies from the individuals they confront. Some key difficulties that affect equity and access are the following:

(a) **Socioeconomic Disadvantage:** Lack of transportation, poverty, and inadequate insurance pose substantial barriers to access to high-quality care.
(b) **Geographical Factors:** In comparison to urban, rural populations generally experience a dearth of healthcare resources and physicians.
(c) **Discrimination:** People experience stigma, discrimination, and even denial of healthcare due to their color, ethnicity, sexual orientation, or gender identity.
(d) **Absence of Sex Education:** Insufficient sexual education heightens susceptibility to deceptive information and hazardous actions.
(e) **Restrictive Laws and Policies:** These can limit access to abortion or restrict the use of contraception.

Disparities in access and equity in reproductive healthcare have serious ramifications: maternal mortality and morbidity increase in the absence of critical services such as contraception. Unequal access to reproductive healthcare is detrimental to gender equality. It becomes a privilege rather than a right to make educated decisions and maintain one's body (Ten Henk and Patrão Neves 2021).

- **Addressing Disparities in Care**

Serious measures should be taken to ensure equity and access to reproductive healthcare for everyone. Any solution to these disparities be multifaceted and cooperative. A few key measures for such a solution can be:

(a) **Strengthening Healthcare Systems:** To assure inexpensive and easy access to essential services, enough resources should be allocated to strengthen healthcare infrastructure and, recruit and train healthcare professionals.
(b) **Sex Education for All:** Such programs can provide adolescents with precise and factual information regarding sexual and reproductive health. They help eliminate social stigmas and damaging preconceptions and allow people to make well-informed choices about their bodies.
(c) **Deconstructing Discrimination:** To create inclusive and welcoming settings for all persons, anti-discrimination legislation should be passed and medical professionals should be trained for cultural sensitivity.

(d) **Community Engagement:** Collaborations with local groups should be made to comprehend their distinct requirements and inclinations. This enables culturally appropriate solutions. Community leaders should be engaged to champion the reproductive health rights of women and girls.

(e) **Advocacy and Activism:** Make underrepresented voices heard, draw attention to the significance of reproductive health justice, and hold governments responsible for preserving reproductive rights.

It will take a significant effort to achieve equity and access to reproductive health. Through collective action, we can create a future in which every person has equitable access to reproductive healthcare and successfully overcome these challenges. An overarching fact is that a strong and empowered populace supports a robust labor market and promotes sustainable development.

8.7 END-OF-LIFE DECISIONS IN REPRODUCTIVE HEALTH

The arena of reproductive healthcare has many ethical concerns. However, ethical concerns related to end-of-life treatment should be handled with extreme sensitivity. In such cases, decisions go beyond the immediate well-being of the individual, and must also take into account the potential consequences for future generations. This presents healthcare professionals with complex ethical dilemmas to grapple with while dealing with terminal illnesses that arise during pregnancy or because of reproductive treatments. In all cases, the basic principles of autonomy, beneficence, and non-maleficence must be at the forefront. This includes protecting patients' autonomy to make informed choices about their reproductive healthcare, ensuring the welfare of present and future generations, and avoiding unnecessary harm (Skuster et al 2023).

When an expecting individual is at risk, deciding whether to continue or terminate the pregnancy requires a careful examination of both the individual's autonomy and the fetus' welfare. The ethical quandary becomes even more complicated when dealing with congenital defects, as offering compassionate reproductive end-of-life care necessitates negotiating moral and emotional difficulties.

Healthcare providers should actively and diplomatically communicate with patients and their families. Comprehensive information about all available options, potential consequences, and the emotional impact of their decisions should be shared. Cultural and religious beliefs generally complicate the delicate situation.

A compassionate and patient-centric approach is crucial for providing meaningful assistance to patients and their families facing tough end-of-life decisions. People should feel empowered to make decisions that align with their values and beliefs. Furthermore, it is critical to acknowledge and address the complex challenges during end-of-life care.

- **Advanced Directives and Living Wills**

Advanced directives and living will play an essential role in shaping the ethical landscape of reproductive health by enabling individuals to convey their preferences and opinions about medical interventions in the event of incapacitation. These legal documents are particularly significant in reproductive health since they guide healthcare decisions including fertility treatments, pregnancy complications, and end-of-life care.

Advanced directives may include specific instructions for reproductive interventions, such as preferences for or against ART, prenatal diagnostics, or whether to continue or terminate pregnancy under particular circumstances. Living wills, on the other hand, allow individuals to express their demands for medical treatments in a broader sense, providing a decision-making framework for when they are unable to communicate.

The ethical considerations surrounding advanced directives and living wills in reproductive health include ensuring that healthcare practitioners respect and follow patients' expressed wishes while balancing autonomy with responsibility for care. Thus, open communication and collaboration between healthcare practitioners, patients, and their families are essential for navigating the many ethical issues that may arise while implementing these directives in the ever-changing and sensitive area of reproductive healthcare.

- **Palliative Care and Dignity in Reproductive Health**

Palliative care needs a compassionate approach that prioritizes the well-being and dignity of individuals facing serious or life-limiting illnesses. This specialized care is meant to alleviate pain, manage symptoms, and address the emotional, spiritual, and social needs of individuals and their families. In the context of reproductive health, palliative care becomes an essential component of holistic patient-centric support. It requires dealing with unique challenges while navigating issues such as fertility, pregnancy complications, and end-of-life decisions.

Ensuring dignity in palliative care requires open communication, autonomy respect, and sensitivity to the cultural and personal beliefs of patients. Healthcare professionals should strive to create an environment where patients feel empowered, heard, and respected. When curative interventions are no longer feasible, the improvement of individuals' quality of life, promoting comfort, and the preservation of their dignity should be the prime focus. This approach not only acknowledges the uniqueness of each patient's experience but also reflects a commitment to compassionate and dignified care.

8.8 SURROGACY AND ASSISTED REPRODUCTION

- **Surrogacy**

It is a condition when a woman carries and gives birth to a child meant for another couple or individual (intended parents), who will then raise the child.

There are two primary kinds:

(a) **Conventional Surrogacy**: In this method, the intended father's sperm fertilizes the surrogate's egg, establishing a genetic link to the kid.

(b) **Gestational Surrogacy**: In this type of surrogacy, the surrogate bears an embryo that was made through IVF using the intended parents' or egg/sperm donors' genetic material; the embryo has no genetic connection to the surrogate.

Surrogacy relies heavily on assisted reproductive procedures including Gamete Intra-Fallopian Transfer (GIFT), Intra-Cytoplasmic Sperm Injection (ICSI), and IVF. By external ova fertilization, these approaches improve the likelihood of successful implantation and subsequent pregnancy (Bayram 2023).

Surrogacy and assisted reproduction have advanced significantly in recent years. They offer couples and people facing infertility or other reproductive challenges with other alternatives. As these procedures need the engagement of a third party throughout the phases of conception and gestation, many ethical, legal, and societal considerations are raised (Dickens and Cook 2017).

- **Ethical Issues in Surrogacy Arrangements**

Significant ethical concerns are raised by surrogacy agreements regarding reproductive rights, autonomy, and the welfare of all parties involved. Several crucial ethical concerns are elaborated upon thereafter:

(a) **Autonomy and Informed Consent:** All relevant stakeholders, including the surrogate, intended parents, and donors, should fully understand the implications of the surrogacy arrangement and provide informed consent.

(b) **Exploitation and Commercialization:** When financial transactions are involved in surrogacy arrangements, there is a potential for exploitation. Ethical consideration should be to ensure that financial agreements are fair, and transparent, and do not coerce individuals into participating.

(c) **Agency and Power Dynamics**: Disparities in power between surrogates and intended parents may compromise the autonomy of the surrogate. Ensuring that the surrogate possesses decision-making autonomy and is not coerced into making specific choices is of the utmost importance.

(d) **Rights and Well-being of the Child:** The rights and well-being of surrogate infants are also ethical concerns. Questions may arise about the child's right to know their genetic origins and the potential impact of the surrogacy process on their identity.

(e) **Selective Reproduction and Genetic Screening:** The ability to examine embryos for specific characteristics or genetic abnormalities gives rise to ethical concerns regarding the potential for selective reproduction, which could perpetuate societal prejudices or stigmatize particular genetic qualities.

(f) **International Surrogacy and Legal Variability:** Ethical concerns may arise due to varying levels of regulation in cross-border surrogacy arrangements. These include inconsistent standards of care, oversight, and protection for all parties involved.

(g) **Emotional and Psychological Well-being:** Surrogacy has the potential to affect the mental and emotional health of all individuals involved. Providing appropriate emotional support, therapy, and mental health services to surrogates, intended parents, and the infant is an ethical imperative.

(h) **Post-birth Relationship Dynamics:** To effectively manage potential emotional issues and expectations, it is crucial to thoroughly contemplate and communicate the postnatal connection among the surrogate, intended parents, and child.

(i) **Cultural and Religious Sensitivity:** Surrogacy may raise cultural or religious concerns that require sensitivity and respect for diverse beliefs and values.

(j) **Regulations and Legal Protection:** The lack of universally accepted surrogacy regulations gives rise to ethical considerations as insufficient legal safeguards for all participants may result in conflicts or exploitation.

Effectively addressing these ethical concerns necessitates thoughtful consideration, candid communication, and a steadfast dedication to safeguarding the welfare, autonomy, and dignity of all individuals involved in surrogacy arrangements.

- **Legal and Ethical Considerations in Third-Party Reproduction**

Legal frameworks about surrogacy differ greatly between nations and even within areas. While some nations allow surrogacy only for altruistic reasons, where payment is restricted to supporting medical and other acceptable costs, others allow commercial surrogacy, in which the surrogate is compensated for her services. Legal contracts that clearly define each party's rights and duties are crucial for settling disputes and guaranteeing a smooth transaction.

8.9 ADOLESCENT REPRODUCTIVE HEALTH

Adolescence, a period of rapid physical and emotional transformation, opens up the intricate world of sexuality and reproduction. Prioritizing adolescent reproductive health yields quantifiable benefits such as reduced unwanted pregnancies, decreased risks of unsafe abortions, and sexually transmitted infections, and enhanced academic achievement. Michaud et al (2023) reported that, in addition, a deeper effect is that young people are empowered to make decisions about their bodies, relationships, and futures with knowledge.

- **Privacy Concerns in Adolescent Health Records**

Adolescents often grapple with issues related to personal identity and sensitive health. Adolescent health records often contain sensitive information about mental and sexual health, substance use, and other personal matters. They may feel particularly vulnerable if their health information is not adequately protected. This makes privacy concerns in adolescent health records a critical issue in today's digital age.

Some of the most significant issues in this field include parental access and involvement, risks associated with data security breaches, stigmatization and discrimination, consent and informed decision-making, third-party access and sharing, legal and ethical challenges, technology-related risks, and a lack of education and awareness.

Hence, a formidable challenge lies in attaining a harmonious equilibrium between the provision of superior healthcare, parental or guardian involvement, and the protection of adolescents' privacy rights. Policymakers and healthcare providers must collaboratively work to establish robust privacy frameworks to safeguard the confidentiality of adolescent health records while ensuring the accessibility of necessary information for appropriate medical care.

- **Consent and Confidentiality for Minors**

In the realm of reproductive healthcare for minors, consent and confidentiality hold significant importance. In this context, a careful equilibrium must be struck between safeguarding the welfare of the young individual and upholding their autonomy. Obtaining informed consent from minors is critical to ensure that they fully comprehend the implications of the care they are receiving.

Confidentiality ensures the privacy of sensitive data about sexual health, contraception, and sexually transmitted infections. It promotes the willingness of minors to seek essential medical attention without apprehension of criticism or revelation to their guardians or parents. This encourages adolescents to make well-informed choices regarding their reproductive health while receiving essential guidance and support.

- **Balancing Parental Involvement with Privacy Rights**

Navigating the delicate balance between parental involvement and the privacy rights of minors in healthcare is a complex challenge. On one hand, parents often play a crucial role in the well-being of their children, and their involvement can be vital for comprehensive care. On the other hand, respecting the autonomy and privacy rights of minors is essential, particularly in sensitive areas such as reproductive healthcare and mental health.

To achieve a suitable balance, recognizing the maturity level of minors and their ability to make choices is extremely important. This enables healthcare providers to assess the intricacy of the healthcare decision and consider applicable legal and ethical standards. Open communication is key to ensuring that minors are well-informed about the implications of their healthcare choices.

Establishing clear and transparent policies and guidelines that define the circumstances under which parental involvement is necessary and when privacy rights take precedence is crucial. It can establish harmonious balance, foster trust among all parties involved, and ultimately promote the best interests of the minor.

8.10 CULTURAL SENSITIVITY

To foster an inclusive and efficient healthcare environment, promoting cultural sensitivity in reproductive healthcare is critical. Identifying and respecting diverse cultural and religious beliefs, practices, and values related to reproductive health is essential to ensure that individuals receive personalized and culturally competent care.

- **Recognizing and Respecting Cultural Differences**

Healthcare professionals must be attuned to unique elements of family structures, religious convictions, and traditional healing practices. This can foster trust and open communication with the patient and lead to more informed decision-making and health outcomes. Furthermore, it aids in the mitigation of disparities in reproductive healthcare accessibility. In all, incorporating cultural sensitivity into reproductive healthcare makes it more patient-centric and equitable (Hiadzi 2023).

- **Cultural Competence in Reproductive Healthcare**

Individuals perceive and make decisions, related to reproductive health, greatly based on their culture. Cultural competence is understanding and respecting these cultural nuances. It can enable healthcare providers to tailor their approach to different cultural practices, beliefs, and values. Cultural competency also includes effective communication considering health literacy levels, and language challenges. It promotes patient autonomy, builds trust, and elevates the overall quality of reproductive healthcare.

- **Ethical Challenges in Cross-Cultural Reproductive Care**

Cross-cultural reproductive care presents a multitude of ethical concerns due to its intersection with beliefs, customs, and healthcare practices. A significant challenge lies in striking a delicate equilibrium between respecting indigenous traditions and upholding universally acknowledged ethical principles. Healthcare professionals are confronted with challenges such as informed consent, which can get complicated by cultural nuances that impact an individual's understanding of medical interventions and treatments. In addition, matters of reproductive autonomy get complex due to familial and societal expectations. An additional ethical predicament arises when discussing reproductive options that potentially run counter to the cultural or religious convictions of a patient, thereby raising apprehensions regarding the constraints of cultural adaptation within the field of medicine.

Some other critical challenges associated with cross-cultural reproductive care are stigmatization and discrimination, access and equity, gender-based norms, traditional and alternative medicine practices, end-of-life and palliative care decisions, surrogate decision-making, genetic counseling and cultural beliefs and cultural competence of healthcare providers.

Striking a balance between cultural sensitivity and upholding ethical standards requires a nuanced and thoughtful approach, necessitating ongoing dialog, education, and the development of guidelines that acknowledge the complexities inherent in cross-cultural reproductive care.

8.11 RESEARCH ETHICS IN REPRODUCTIVE HEALTH

Research on reproductive health can greatly enhance the quality of life, develop medicine, and expand medical understanding. This will empower people with the knowledge needed to make educated decisions about their bodies and families. However, given the sensitive nature of this discipline, any research in this area must adhere to ethical principles to protect vulnerable communities, groups, and people. Maintaining the greatest ethical standards and producing insightful knowledge are two difficult dances to master.

- **Ethical Conduct in Reproductive Health Research**

Participants' autonomy should be of prime importance in any medical research. Participants should be free to leave the study at any time and without facing any repercussions. This highlights their autonomy in making decisions and honors their changing desires.

Participation in the community, beyond individuals, is essential. To ensure that research is in line with the community's needs and values, research programs should actively involve the communities they will affect. Singh et al (2019) stated that the change can be done by working with them and getting their feedback.

Building trust and ensuring long-term sustainability require equitable sharing of research results with study participants and the communities to which they belong, in addition to being a moral need.

- **Informed Consent in Clinical Trials**

The informed consent principle is central to research ethics. Before giving their consent, participants must be thoroughly informed of the goals, dangers, and advantages of engaging in research. Consent should be based on open and transparent communication, cultural sensitivities, and assurance that participants are capable of making educated judgments.

- **Protecting Vulnerable Populations in Research**

In addition to informed consent, vulnerability needs to be given special consideration in studies on reproductive health. The vulnerability of women, children, and marginalized communities might be heightened by societal stigma, power disparities, or restricted access to healthcare resources. Essential measures to prevent vulnerable groups' exploitation are preserving anonymity, guaranteeing confidentiality, and giving priority to low risk.

8.12 EDUCATIONAL AND SUPPORTIVE RESOURCES

Educational and supportive resources on the ethics of reproductive health are of the utmost importance. These provide lawmakers, medical practitioners, and researchers with the information and skills required to negotiate the complex ethical terrain of reproductive health research. These tools promote the protection of participants' rights and cultivate a responsible culture that maintains ethical standards in research (MacKlin 1990).

The following are some crucial educational and support programs:

(a) **Ethics Training Programs:** Academic institutions, scientific bodies, and trade associations often offer courses on essential principles of research ethics. Such courses cover equity, confidentiality, informed consent, cultural sensitivity, etc. These training programs often incorporate case studies.

(b) **Guidelines and Frameworks:** Prominent institutions such as the World Health Organization (WHO), the International Conference on Harmonisation (ICH), and local health authorities furnish specifically designed comprehensive reference materials, rigorous standards, and ethical frameworks for the investigation of reproductive health.

(c) **Ethics Committees and Review Boards:** Institutions often establish IRBs or ethics committees to meticulously review and approve reproductive health research protocols. These committees guarantee regulatory compliance, provide ongoing support during the research process, and advise on ethical concerns. Researchers greatly benefited when these committees were engaged in the early first stages of the study.

(d) **Online Platforms and Courses:** Various institutions and websites provide specialized tools, webinars, and courses tailored for academics and medical professionals. Such platforms provide handy educational information on many subjects, including the ethical quandaries unique to reproductive health research and possible answers.

(e) **Professional Associations and Conferences:** Associations often host symposiums, seminars, and conferences that focus on ethical issues related to medical research, reproductive health, and bioethics. In addition to networking, these platforms enable exchanging stories, and remain up to date on the latest ethical principles and best practices. Participants may gain valuable perspectives on intricate ethical quandaries from experts.

(f) **Publications and Journals:** Specialized scholarly publications should be dedicated to bioethics and reproductive health, and disseminate research findings, analyses, and commentary. Regularly reading and actively interacting with this continuously growing body of information helps academics stay up to date on important ethical issues and creative solutions.

(g) **Patient Advocacy Groups:** Advocacy organizations and community organizations often promote ethical research protocols. Collaborating with them may offer researchers important perspectives and improve the pertinence and inclusivity of studies.

(h) **Mentorship Programs:** Programs designed to facilitate emerging professionals by pairing with established academics can advance knowledge and expertise in the domain of reproductive health ethics. It offers a safe space for mentors and mentees to share important guidance and the exploration of ethical issues.

REFERENCES

Bayram DV (2023) Reproductive health and ethical problems in women's health. In: *Midwifery – New Perspectives and Challenges* [Working Title]. IntechOpen.

Brezina PR, Zhao Y (2012) The ethical, legal, and social issues impacted by modern assisted reproductive technologies. *Obstetrics and Gynecology International* 2012:1–7. https://doi.org/10.1155/2012/686253

Brown AEC, Hobart TR, Morrow CB (eds) (2019) *Bioethics, Public Health, and the Social Sciences for the Medical Professions: An Integrated, Case-Based Approach*. Springer International Publishing, Cham.

Christianson M, Lehn S, Velandia M (2022) The advancement of a gender ethics protocol to uncover gender ethical dilemmas in midwifery: a preliminary theory model. *Reproductive Health* 19:211. https://doi.org/10.1186/s12978-022-01515-6

Cook RJ, Dickens BM, Fathalla MF (2003) *Reproductive Health and Human Rights*. Oxford University Press, Oxford.

Davis MF (2022) The state of abortion rights in the US. *International Journal of Gynecology & Obstetrics* 159:324–329. https://doi.org/10.1002/ijgo.14392

Dickens BM, Cook RJ (2017) Legal and ethical factors in sexual and reproductive health. In: Edozien LC, O'Brien PMS (eds) *Biopsychosocial Factors in Obstetrics and Gynaecology*. Cambridge University Press, Cambridge, pp 209–218.

Hiadzi RA, Woodward B, Akrong GB (2023) Ethical issues surrounding the use of assisted reproductive technologies in Ghana: an analysis of the experiences of clients and service providers. *Heliyon* 9:e13767. https://doi.org/10.1016/j.heliyon.2023.e13767

MacKlin R (1990) Ethics and human reproduction: international perspectives. *Social Problems* 37:38–50. https://doi.org/10.2307/800793

Michaud PA, Takeuchi YL, Mazur A, et al (2023) How to approach and take care of minor adolescents whose situations raise ethical dilemmas? A position paper of the European Academy of Pediatrics. *Frontiers in Pediatrics* 11:1120324. https://doi.org/10.3389/fped.2023.1120324

Muhunthan K, Arulkumaran S (2014) Medical Ethics and Human Rights in Reproductive Health. *Nepal Journal of Obstetrics and Gynaecology* 9:5–7. https://doi.org/10.3126/njog.v9i1.11177

Premkumar A (2018) Reproduction, inequality, and technology: the face of global reproductive health ethics in the twenty-first century. *AMA Journal of Ethics* 20:224–227. https://doi.org/10.1001/journalofethics.2018.20.3.fred1-1803

Saenz C, Cheah PY, Van Der Graaf R, et al (2017) Ethics, regulation, and beyond: the landscape of research with pregnant women. *Reproductive Health* 14:173. https://doi.org/10.1186/s12978-017-0421-3

Schenker JG, Eisenberg VH (1997) Ethical issues relating to reproduction control and women's health. *International Journal of Gynecology & Obstetrics* 58:167–176. https://doi.org/10.1016/S0020-7292(97)02866-X

Schweikart S (2018) AMA Code of Medical Ethics' opinions related to global reproductive health. *AMA Journal of Ethics* 20:247–252. https://doi.org/10.1001/journalofethics.2018.20.3.coet1-1803

Shaw D (2004) History of the FIGO Committee for Women's Sexual and Reproductive Rights. *International Journal of Gynecology & Obstetrics* 86:294–316. https://doi.org/10.1016/j.ijgo.2004.05.002

Singh JA, Siddiqi M, Parameshwar P, Chandra-Mouli V (2019) World Health Organization Guidance on ethical considerations in planning and reviewing research studies on sexual and reproductive health in adolescents. *Journal of Adolescent Health* 64:427–429. https://doi.org/10.1016/j.jadohealth.2019.01.008

Skuster P, Moseson H, Perritt J (2023) Self-managed abortion: aligning law and policy with medical evidence. *International Journal of Gynecology & Obstetrics* 160:720–725. https://doi.org/10.1002/ijgo.14607

Ten Have H, Patrão Neves MDC (2021) Reproductive ethics. In: *Dictionary of Global Bioethics*. Springer International Publishing, Cham, pp 891–891

CHAPTER 9

Challenges in Reproductive Health

Amita Sharma, Geeta Pandey, Veni Venu, Devika Asokan, Sreemoyee Chatterjee, Harshita Bhargava, Barkha Khilwani, Abdul S. Ansari, Nirmal K. Lohiya, Pushpendra Singh, and Prashanth N. Suravajhala

9.1 INTRODUCTION

The World Health Organization (WHO)/United Nations Population Fund Report defines reproductive health (RH) as full physical, mental, and social well-being throughout the reproductive process (UNPF, 1994). Over the years, several reports stated that RH is a major concern in the realm of women's health with impaired developmental effects (Woodruff et al., 2008). COVID-19 has taken leaps and bounds in women's health and efforts have been on the rise to bring up various policies related to this (Berg et al., 2022). The integrated strategy connects several factors, including avoiding pelvic infections that cause infertility, assuring baby survival via effective prenatal and antenatal care, and controlling sexually transmitted illnesses like HIV. The RH is not restricted to mothers or women of childbearing age; it also addresses the unique needs of adolescent and adult women. In women, cervical cancer (CC) occurs because of the persistent infection of human papillomavirus (HPV) wherein screening tests confirm whether or not the virus exists in the cervix bounded with genomic risk factors (Ramachandran and Dörk, 2021). One of the major goals of RH is early detection and screening in women which remains a major challenge (Basoya and Anjankar, 2022). On the other hand, men's RH is with significant health risk posed by the increasing incidence of male-related malignancies, infertility, and reproductive disorders, which have shown a notable rise in recent years. Specifically, impaired sperm characteristics are found in people with testicular or prostate cancer (PCa), which are similar to those previously reported in patients who were

DOI: 10.1201/9781003487548-9

infertile or subfertile before receiving general therapy. Men who have regular reproductive testing have an increased likelihood of early diagnosis for their risk of developing cancer, which exacerbates this association (Tvrda et al., 2015). Infertility and PCa, often characterized by a multifactorial origin, are intricately linked to both genetic and environmental influences. Approximately 50% of infertility cases are attributed to genetic factors, a connection extensively validated in numerous animal model studies showcasing associations between infertility and singular or multiple gene defects (Rodrigues et al., 2020).

While lifestyle diseases and cancers impact RH, there are also environmental factors like air and/or water pollution, and exposure to certain toxins like lead, arsenic, and more, which contribute significantly (Oczkowski et al., 2021). Lifestyle factors including dietary habits such as excessive calorie or fat intake, cholesterol-rich diets, increased consumption of red meat, fruits, fish, and dairy products, dietary calcium, as well as alcohol consumption, lack of physical activity, and many more have been suggested as factors elevating the risk for cancers of RH including PCa cancer. Understanding the interplay between genetic predispositions and environmental influences is crucial for unraveling their complexities, and facilitating improved risk assessment, prevention, and targeted therapeutic interventions. Considerable evidence from epidemiological, genetic, histological, and molecular studies highlights the significant role of RH-associated ailments ensuing the onset of chronic inflammation. For example, in PCa, infectious agents potentially affect carcinogenesis through several mechanisms: (a) integration of viral oncogenes into the carrier's genome, (b) suppression of tumor suppressor genes, (c) promotion of proliferative signals, and (d) dampening of the immune surveillance system. While chronic prostatitis stemming from sexually transmitted diseases (STDs) correlates with heightened PCa risk and compromised treatment outcomes, as of now, no specific infectious agent has been definitively linked as a causative factor of the disease (Gann, 2002). However, transferring these discoveries into clinical trials has been proven difficult. Through personalized medicine and the completion of the human genome project have made great advancements, only a few genes and genetic alterations have been proven to be directly associated with primary infertility. While significant studies have been made in enhancing global RH over the past two decades through focused policies and programs, challenges persist in effectively implementing policies, especially for underserved populations. The Asian and Pacific regions have seen improvements in sexual and RH, yet barriers hinder the translation of policies into impactful programs, particularly for the marginalized. Despite notable progress in sexual and RH indicators in the WHO South-East Asia Region between 2000–2017 and 1990–2018, such as a 57% reduction in maternal deaths and a 60% decrease in newborn mortality, PCa remains unaddressed in these studies. Efforts in the region aim to further decrease mortality rates and ensure universal access to sexual and RH services (Acuin et al., 2011). This chapter coalesces challenges on prevention and diagnostics related to RH-associated anomalies like CC and PCa within the broad context of systems biology with a special emphasis on genomics. The goal is to discuss improved drugs, therapies, and measures for enhancing the well-being of the reproductive system. We finally provide grand challenges in the realm of RH measures.

9.2 GENOMIC EVOLUTION OF REPRODUCTIVE HEALTH

Carcinogenesis is the transformation of normal cells into cancer cells, a complex process characterized by various genetic and epigenetic changes. Oncogenesis in this context arises from a multitude of genomic and epigenomic alterations within these cells. Globally, 25% of newly diagnosed instances of male cancer are PCa. The prevalence of PCa rises sharply with age, from 166 to 800 cases per 100,000 males aged 55–59 to 75–79, respectively (Ntekim et al., 2023). Generally, genomic instability is considered a key component of human cancer, including oncogene activation and tumor suppressor gene inactivation which may drive abnormal cell growth. Oncogenes cause aberrant cell proliferation due to genetic changes that either boost gene expression or cause uncontrolled activation of the oncogene-encoded proteins. The modifications in PCa are mostly caused by structural genomic rearrangements over time, which lead to gene amalgamations, amplifications, and omissions (Li and Ralph, 2019). It often begins with genetic mutations in normal prostate cells, leading to the initiation of cancerous growth. Common early mutations involve genes such as PTEN, TP53, and TMPRSS2-ERG rearrangements. Loss of the tumor suppressor gene PTEN is a frequent event in PCa, leading to increased cell proliferation and survival. Thus, activation of proto-oncogenes and suppression of tumor suppressor genes within the mitogen-activated protein kinases (RAS/MAPK), STAT3, and PI3K/AKT signaling pathways is correlated with the onset, development, and advancement of PCa (Rybak et al., 2015). Normal genes known as proto-oncogenes influence normal cell growth and proliferation, but if their expression is changed, they may also play a role in the formation of cancer. Deletions in genes such as NKX3-1, PTEN, BRCA2, and RB1 are common in initial instances, but metastatic cases exhibit chromosomal augmentations affecting androgen receptor pathway genes and the MYC oncogene. At least 20 mutations affect protein function in PCa, serving as a key tumor suppressor gene even as castration-resistant (hormone-refractory) prostate cancer (CRPC), which accounts for the majority of PCa deaths, is a result of androgen-dependent carcinomas that advance through oncogenic signaling pathways during the carcinogenesis of the prostate (Rybak et al., 2015). In hereditary cases, PCa shares a close link with hereditary breast cancer, especially through BRCA1 and BRCA2 mutations. Carriers of mutations exhibit a 3.8-fold (BRCA1) or 8.6-fold (BRCA2) increase in PCa risk. BRCA2 mutations indicate aggressive disease with higher mortality rates, affecting around 5% of advancing PCa cases. The hereditary element is exemplified by the BRCA2 gene on chromosome 17p, associated with DNA cross-link repair enzymes. Specific genes like Arg293X and Asp175Ty have been identified in familial PCa, while HOXB13 G84E polymorphisms suggest elevated risk which demonstrates the critical role that genetic assessment plays in assessing and managing cancer risk further affecting RH (Huang and Cai, 2014).

The different stages of PCa are shown in Figure 9.1.

Some of the major treatment options include active surveillance measures, chemotherapy, and radiation therapy combined with hormonal therapy which has been shown to enhance survival rates, for example, stereotactic radiotherapy for low/intermediate

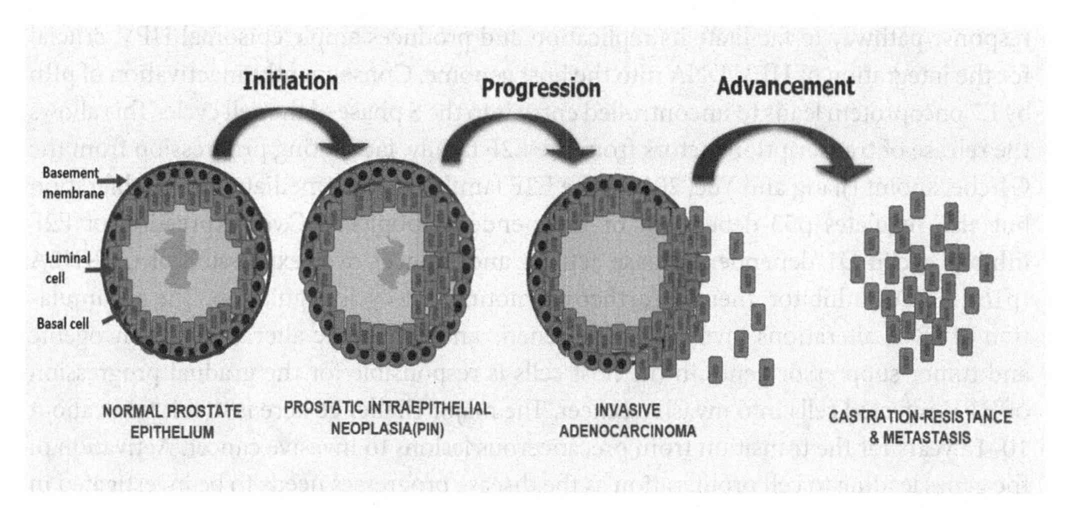

FIGURE 9.1 Prostate Cancer and Stages.

PCa risk whereas brachytherapy, an alternative radiation therapy method, offers an advantageous treatment option for patients facing challenges such as transportation issues that may complicate standard external beam therapy (Mayer et al., 2024). In the recent past, radical prostatectomy has been used which offers the greatest potential for a definitive cure for localized PCa. Cryotherapy, on the other hand, employs extremely low temperatures to freeze and eliminate cancer cells as it is used for overcoming high incidences of erectile dysfunction.

Cervical cancer is caused by a large number of squamous cells deteriorating with about 90% of CC cases being associated with the HPV infection. While it is among the few completely preventable cancers, early detection is now possible by self-sampling, home tests, and simple procedures that have significantly reduced the mortality rate (Poondla et al., 2021). Over the years, paramount studies on early diagnosis, and preventive measures along with raising awareness among adolescent girls and also women are promising and cost-effective. The HPV oncoproteins, predominantly E6 and E7, play a crucial role in altering the function of host cells. They are known to overexpress E6 and E7 oncoproteins to disrupt the normal functioning of tumor suppressor genes within the host. Once integrated, these viral proteins initiate damage to the host cells. Host cells, in response, employ a specific mechanism to repair the DNA damage inflicted by the viral proteins through DNA damage response pathways. Repairing the damaged DNA is essential for mitigating cell cycle checkpoints, allowing cells to resume division. However, in cases where the cells fail to repair DNA damage, apoptosis, or programmed cell death, occurs. In HPV-related cancers, E6/E7 viral proteins interfere with cell-cycle checkpoint control by inhibiting cyclin-dependent kinase (CDK) inhibitors (such as p21, p27, and p16) and degrading key regulatory proteins like p53 and retinoblastoma (pRb) (Yim and Park, 2005). The degradation of p53 induced by E6 oncoprotein prevents apoptosis, enabling continuous cell replication, while the degradation of pRb by E7 oncoprotein prompts unscheduled entry into the S phase of the cell cycle, promoting cell proliferation. HPV exploits this damage response pathway to facilitate its replication and produces ample episomal HPV, crucial for the integration of HPV DNA into the host genome. Consequently, inactivation of pRb by E7 oncoprotein leads to uncontrolled entry into the S phase of the cell cycle. This allows the release of transcription factors from the E2F family, facilitating progression from the G1 checkpoint (Jiang and Yue, 2014). The E2F family not only mediates cell proliferation but also regulates p53-dependent or independent apoptosis. Over-expression of E2F inhibits cyclin D1-dependent kinase activity and induces over-expression of CDKN2A (p16), a CDK inhibitor, thereby further promoting cell-cycle regulation. The accumulation of DNA alterations involving both genetic and epigenetic alterations in oncogenic and tumor suppressor genes in the host cells is responsible for the gradual progression of HPV-infected cells into invasive cancer. The major challenge here is that it takes about 10–12 years for the transition from precancerous lesions to invasive cancer. Activation of the genes leading to cell proliferation as the disease progresses needs to be investigated in finer detail to design effective interventions.

Self-sampling for HPV testing presents various advantages, particularly in low- and middle-income countries (LMICs), where its simplicity and convenience make it a valuable tool. Not only does self-sampling offer ease of use, but it also provides physical and emotional comfort to individuals undergoing testing. However, a major diagnostic challenge is to ensure the accuracy and efficacy of the self-test.

In the treatment regime, the additional factor determining whether the preservation of fertility is both desired and feasible stands as a pivotal consideration. The decision-making process must encompass the patient's preferences, overall health status, existing risk factors, menopausal stage, and current life circumstances. These factors collectively inform the appropriate course of action regarding fertility preservation. The lymph node status of the patient is among the foremost prognostic indicators of CC. Decisions regarding tumor stage determination, prognosis assessment, and subsequent therapeutic strategies heavily rely on intraoperative evaluation of lymph node status. Studies indicate that preoperative imaging modalities such as computed tomography (CT), magnetic resonance imaging (MRI), or positron emission tomography-computed tomography (PET-CT) are less reliable compared to surgical methods for detecting lymph node metastases. Therefore, intraoperative assessment remains crucial for accurate staging and treatment planning (Thoenissen et al., 2023).

The genetic modifications associated with the advancement of PCa are shown in Figure 9.2.

As we say prevention is better than cure, vaccines if available are considered as a boon in preventing diseases like cancer. The HPV vaccine is a crucial tool in preventing HPV infections and related cancers. It is recommended for both males and females to protect against several types of HPV that can lead to cervical, anal, penile, and throat cancers. Studies have consistently demonstrated the safety and effectiveness of HPV vaccination in reducing HPV-related diseases (Markowitz et al., 2024).

Similarly, in a classical radical hysterectomy, the procedure typically entails the total removal of the cardinal ligament, which includes the excision of the pelvic splanchnic

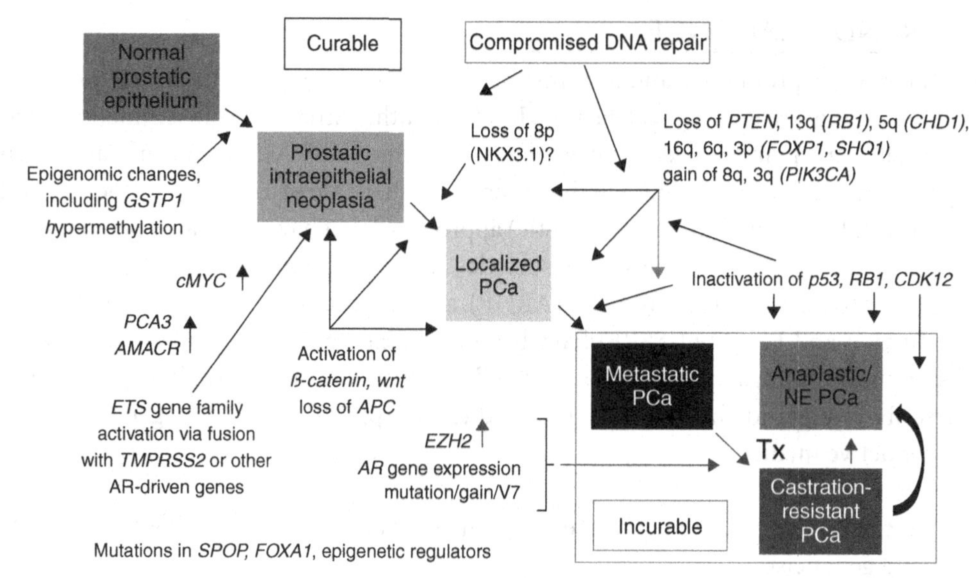

FIGURE 9.2 Genomic Alterations in Prostate Cancer Progression.

nerves. In consideration of the fact that 40% of CC patients are of reproductive age, the preservation of fertility becomes a significant concern, provided the patient expresses a desire to retain fertility and when such preservation aligns with oncological principles. According to guidelines from the National Comprehensive Cancer Network (NCCN), the European Society of Gynecological Oncology (ESGO), the European Society for Radiotherapy and Oncology (ESTRO), the European Society of Pathology (ESP), as well as German guidelines, fertility-sparing treatment may be offered to patients diagnosed with stage IA1 to IB1 squamous cell carcinoma and adenocarcinoma of the cervix (ESGO guidelines, 2014).

Sexual contact is often the means by which sexually transmitted infections (STIs) are contracted. Usually, sexual contact—including body fluids or skin contact during vaginal, oral, or anal sex—is how the bacteria, viruses, or parasites implicated are spread (Fasciana et al., 2022). The WHO reports that over a million STIs are diagnosed globally each day (Tuddenham et al., 2022). The following four STDs were among the 376 million new infections recorded in 2016: syphilis (6.3 million), gonorrhea (86.9 million), chlamydia (127.2 million), and trichomoniasis (156 million) (Rowley et al., 2019). STDs have a serious negative impact on the health of those who are afflicted and can lead to several secondary illnesses and problems, including CC, infertility, pregnancy-related problems like fetal mortality, increased risk of contracting HIV, and decreased quality of life as a result of psychological and social factors (Korenromp et al., 2019). While Herpes simplex virus type 1 (HSV-1) and type 2 (HSV-2) are extremely common, with global prevalence rates of around 67% and 13%, respectively. The challenge here is to look into neonatal infections that can affect the skin, eyes, or mouth, damaging the central nervous system or several organs.

9.3 GRAND CHALLENGES

Over the years, reproductive and genomic medicine have allowed translational practices in the emerging areas of sexual health, Infant health, intrauterine insemination, and men's health largely in the areas of cancer adenocarcinoma (CA) prostate and androgen receptor signaling to name a few. On the other hand, there are prognostic, diagnostic, and theranostic (also referred to as theragnostic) approaches that have facilitated personalized medicine to the fore. Many professional societies and associations such as the Association of Reproductive Health Professionals (ARHP) and the Indian Society for the Study of Reproduction and Fertility (ISSRF) have been instrumental in developing a framework to disseminate findings in these areas. Given the aforementioned myriad of applications, we describe five grand challenges on the road to the application of these genetic testing in reproductive medicine:

(a) Societal awareness for genetic testing, which would bridge the gap between society and geneticists
(b) Cataloging candidate mutations for screening to develop a 'One Reproductive Health' panel
(c) Therapeutic targets for bringing precision medicine
(d) Health, Hygiene, and Habit (3H)
(e) Bringing high-throughput technologies into practice.

9.3.1 Societal Awareness for Genetic Testing

Hereditary disorders can be prevented, treated, or detected early with genetic testing. It is known that 5–10% of cancers, including breast, prostate, and colorectal cancers, have hereditary components. There is a lack of awareness about the importance of genetic testing, leading to a rise in hereditary disorders. Genetic testing often evokes fear in families and individuals, not only in RH but also in various diseases. Sickle cell disease (SCD), a group of inherited red blood cell disorders, exemplifies this concern. SCD, an autosomal recessive genetic condition, leads to red blood cells with a distinctive sickle shape. It predominantly affects individuals of African, Mediterranean, Middle Eastern, Indian, Caribbean, and Central and South American descent. Comprehensive follow-up programs following the detection of SCD traits in newborn screening are inadequate. The uptake of genetic counseling and follow-up services has proven to be challenging (Long et al., 2011) despite African-Americans' positive views and reservations.

The 3Hs and the grand challenges in genome-based reproductive medicine are depicted in Figure 9.3.

Mutational Cataloging and Therapeutic Targets about genetic testing (Gustafson et al., 2007) reported a considerable increase in the perceived benefits of genetic testing for SCD. This emphasizes how crucial it is to encourage genetic testing and counseling, particularly in the area of RH, to narrow the gap between geneticists and the general public. By boosting public awareness, we can promote informed decision-making, reduce stigma,

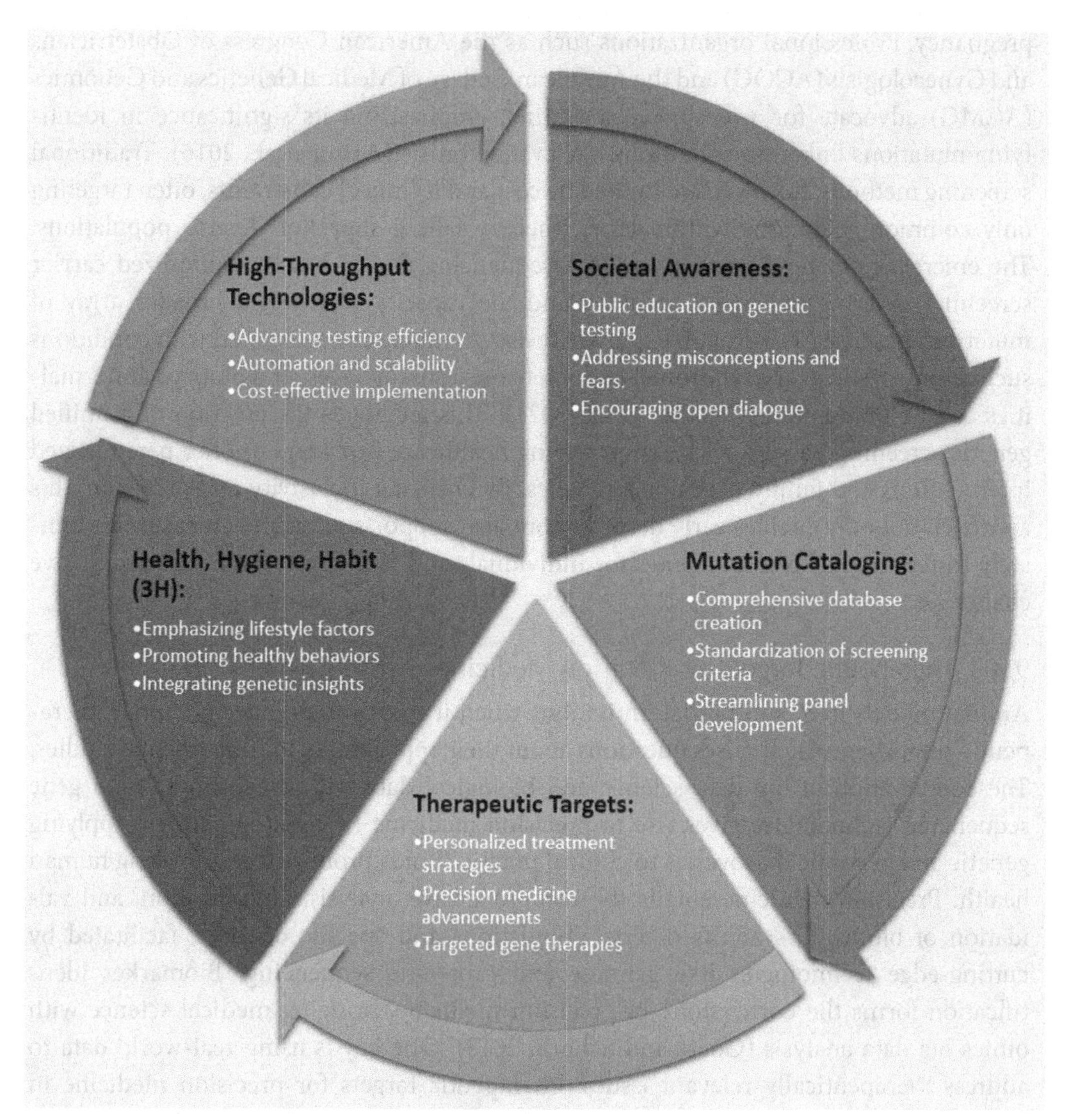

FIGURE 9.3 Grand Challenges in Genome-based Reproductive Medicine, viz. 3Hs: Health, Hygiene, and Habit; High-Throughput Technologies; Societal Awareness.

and provide equal access to genetic services, providing an encouraging environment in which people can make educated decisions about their RH.

9.3.2 Cataloging Candidate Mutations for Screening

Effective cataloging of candidate mutations is pivotal in constructing a comprehensive RH panel. This process involves a meticulous compilation of genetic variations associated with a spectrum of reproductive disorders, encompassing fertility issues, pregnancy complications, and offspring health concerns. Carrier screening, coupled with genetic counseling, has proven to substantially diminish the prevalence of recessive genetic disorders by informing individuals about their reproductive risks before or during pregnancy. Professional organizations such as the American Congress of Obstetricians and Gynecologists (ACOG) and the American College of Medical Genetics and Genomics (ACMG) advocate for widespread screening, emphasizing its significance in identifying mutations linked to conditions like cystic fibrosis (Azimi et al., 2016). Traditional screening methods, however, are limited by cost and technical constraints, often targeting only common mutations within genes, thereby falling short for diverse populations. The emergence of next-generation DNA sequencing (NGS) has revolutionized carrier screening, offering enhanced accuracy and the capacity to assess a broader array of mutations per disease. Through extensive research, mutations associated with conditions such as polycystic ovary syndrome, endometriosis, miscarriages, and embryo abnormalities can be pinpointed (Capalbo et al., 2021). This facilitates the creation of a unified genetic screening panel for RH, empowering healthcare providers to offer personalized interventions and family planning strategies. By ensuring inclusivity and precision, this approach not only enables early detection but also supports preventive measures, ultimately fostering improved outcomes for individuals and families navigating reproductive challenges.

9.3.3 Therapeutic Targets for Precision Medicine

Animal models have substantial limitations when it comes to predicting human therapeutic responses, which raises questions about their applicability in fundamental studies. The combination of big data science and biological data with breakthroughs in gene sequencing technologies gives rise to precision medicine (Song et al., 2020). Applying genetic and genomic discoveries to clinical practice holds promise for improving human health. Precision medicine entails the comprehensive analysis, identification, and validation of biomarkers across diverse populations and specific diseases, facilitated by cutting-edge technologies like genome and proteome sequencing. Biomarker identification forms the cornerstone of precision medicine, bridging medical science with omics big data analysis (Goetz and Schork, 2018). One key is using real-world data to address therapeutically relevant issues. Therapeutic targets for precision medicine in RH include immune modulation for conditions like recurrent miscarriages, genetic and molecular interventions for infertility and endometriosis, epigenetic modifications, microbiome manipulation, and personalized lifestyle interventions. These strategies offer tailored approaches to enhance RH care, promising improved outcomes for individuals and couples facing reproductive challenges. There is a need for therapeutic target-based approaches to imbibe these practices in RH.

9.3.4 Health, Hygiene, and Habit

Hygiene behaviors are critical for long-term health and have a significant impact on RH processes in humans. According to the WHO, RH conditions such as STDs are a serious global health concern, with an increasing number of cases occurring worldwide. Drug resistance, environmental variables that facilitate transmission, and enduring stigma are some of the barriers that limit success in the fight against STIs. Infections linked to STDs are prevalent in adolescents. They need to understand the complexities of STD prevention, which requires thorough knowledge of RH (Mahobia and Choudhari, 2022). Neglecting personal hygiene, such as using polluted water or changing underwear infrequently, can result in illnesses that facilitate STD transmission (Mariani et al., 2021). Young women typically experience issues with menstrual hygiene (Pokhrel et al., 2021). These issues still exist, particularly for girls of school age and those who reside in low-income areas. These challenges originate from traditional attitudes, limited access to knowledge, and a scarcity of sanitary items (Parent et al., 2022).

Reproductive diseases in adults are frequently linked to sexual behavior in addition to poor hygiene. Therefore, to comprehend RH, one must be knowledgeable about the reproductive system, health practices, contraception, and STD prevention (Lindberg et al., 2020). Encouraging awareness of personal hygiene's vital role in preserving general health is part of the promotion of hygiene. RH problems can also occur in professional contexts, such as laboratory work. Laboratory personnel are protected against RH concerns using conventional hygiene standards (Mizan, 2016). Personal hygiene is a concern in RH in this instance also. To achieve this, hazardous materials must be replaced, ventilation system effectiveness must be guaranteed, and engineering solutions must be used instead of personal protective equipment. Thus, integrating 3H principles into RH education, services, and policies is essential for optimal RH outcomes. It is important to emphasize lifestyle choices including abstaining from drugs, having safe sexual relations, and developing healthy mental health practices. Reducing socioeconomic gaps guarantees fair access to reproductive healthcare, promoting health among various groups.

9.3.5 Implementing High-throughput Technologies

There is tremendous promise for improving diagnosis, treatment, and research in the field of reproductive disorders and RH by implementing high-throughput technologies. Personalized medicine approaches are made possible by these technologies, which include high-resolution imaging, microarrays, and NGS. These technologies allow for the quick analysis of genetic, epigenetic, and proteomic data. In assisted reproduction technology, there are many examples of new methods and tools being introduced into the therapeutic context without the required research and development or evidence-based medicine to reinforce the procedure's safety and benefits for the patient. Examples include blastocyst transfer, vitrification, in vitro maturation, assisted hatching, and preimplantation genetic screening. It is common for international standards to be established without adequate validation for the makeup of culture medium, stimulation schedules, and laboratory procedures (Clarke et al., 2012; Harper et al., 2012). High-throughput technology can help detect genetic abnormalities associated with infertility in reproductive illnesses, allowing for customized therapies. These technologies also improve screening for sexually transmitted illnesses, resulting in earlier discovery and treatment. High-throughput technologies such as NGS and microarray analysis allow for thorough genomic profiling in the case of reproductive malignancies (Dawes et al., 2019), such as PCa, finding genetic abnormalities linked with cancer development, progression, and response to treatment. Precision medicine techniques are made possible by these technologies, which enable individualized treatment plans based on the distinct genetic profiles of each patient. Even the identification of biomarkers, which can enhance PCa patients' early detection and prognostic evaluation, is facilitated by high-throughput approaches (Böttcher, 2016). Healthcare practitioners can improve accuracy, efficacy, and efficiency in addressing RH issues by incorporating these technologies into clinical practice and research projects.

Several significant obstacles must be overcome before genetic testing may be used in reproductive medicine. These include raising public awareness, organizing potential mutations, finding therapeutic targets, encouraging good health and hygiene habits, and putting high-throughput technology into practice. To overcome these obstacles, stakeholders must work together, make investments in education and research, and remain dedicated to the advancement of precision medicine techniques. Personalized care and better RH outcomes will be made possible by addressing these issues.

9.4 GENETIC APPROACH

Genetic screening is used to identify individuals who have genetic variations or mutations that are associated with a particular disorder. The goal of genetic screening is to provide information to individuals regarding their health and the health of their offspring. Whereas genetic screening may be useful to plan for periodic evaluation, implementation of preventive strategies, or initiation of therapeutic interventions, decreasing fertility is one of the major contributors to more people adopting genetic screening/counseling.

Today, infertility is a major concern globally affecting millions of people of reproductive age. Worldwide approximately 48 million couples and 186 million individuals have infertility. In 65% of cases diagnosis of infertility includes biochemical and instrumental analyses, the remaining 35% of cases, which are undiagnosed, require genetic tests to be performed. Infertility has been associated with 15% of genetic disorders; genetic and nongenetic causes of infertility show a similar pattern of clinical symptoms. Thus, to manage this ever-increasing infertility problem, clinical diagnosis should be supported and confirmed through genetic evaluation. Genetic tests serve as a valuable tool in pinpointing potential genetic disorders that could be inherited by offspring. Through preconception screening, prospective parents gain insight into their reproductive risks before conception, helping them to make informed decisions about family planning. This proactive approach not only aids in understanding potential hereditary health concerns but also facilitates early interventions and personalized medical care, ultimately paving the way for healthier outcomes for future generations. Normally, gametes with genetic or chromosomal defects sometimes have difficulty reproducing. However, developments in assisted reproductive technology (ART) have completely changed this environment by providing ways to get beyond these impediments. Genetic testing, including carrier screening, preimplantation, and prenatal diagnosis, is critical for monitoring the possible transfer of specific defects in genes to adolescents.

In the field of prenatal screening and testing, antenatal diagnosis presents an interesting new horizon for molecular diagnostics. This specialized application involves the detection of genetic abnormalities, diseases, or other conditions in the unborn fetus, offering invaluable insights into the health status of the developing baby prior to birth. Preimplantation genetic testing (PGT) and prenatal diagnosis (PND) are the current diagnostic approaches used for couples who may be at risk of passing on inherited disorders. The scheduling, sample techniques, and laboratory protocols of these treatments vary, even though they all aim to uncover genetic abnormalities. In addition, molecular biology methods used in PGT, in conjunction with conventional laboratory studies, have significantly improved the efficacy of ART operations. These innovations not only accelerate procedures, but also help them succeed by lowering time, effort, and costs.

9.5 SINE DIE FOR GENETIC TESTING?

There are many different kinds of genetic tests (Figure 9.4). There is no single genetic test that can detect all genetic conditions. The approach to genetic testing is individualized based on medical and family history and conditions being tested for. For instance, single-gene tests focus on identifying variations within a specific gene when symptoms of a particular condition or syndrome are present, such as Duchenne muscular dystrophy (DMD) or SCD. They are particularly useful in cases where a known genetic mutation runs in a family. Conversely, panel genetic testing examines multiple genes in one comprehensive test. These tests are typically categorized based on various medical concerns, like low muscle tone, short stature, or epilepsy. Also, panel genetic tests can target genes associated with increased susceptibility to specific types of cancer, such as breast or colorectal cancer. This approach offers a broader scope for identifying potential genetic factors contributing to a range of health conditions. There is an inherent need for large-scale genetic or genomic testing: There are two different kinds of large-scale genetic tests. Exome sequencing looks at all the genes in the DNA (whole exome) or just the genes that are related to medical conditions (clinical exome). Large-scale genetic testing may identify

Carrier Screening:
- Identifies genetic mutations.
- Example: Testing for BRCA mutations in breast cancer risk assessment.

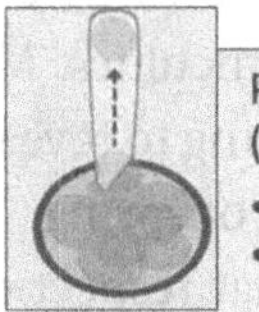

Preimplantation Genetic Testing (PGT):
- Evaluates embryos before implantation.
- Example: Screening embryos for genetic abnormalities in IVF procedures.

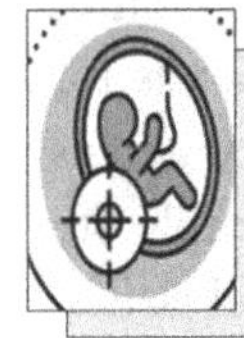

Prenatal Testing:
- Detects genetic disorders in fetuses.
- Example: Amniocentesis for detecting chromosomal abnormalities during pregnancy.

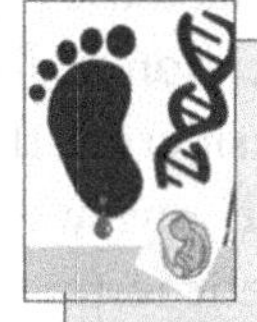

Newborn Screening:
- Identifies genetic conditions in newborns.
- Example: Testing for phenylketonuria (PKU) shortly after birth.

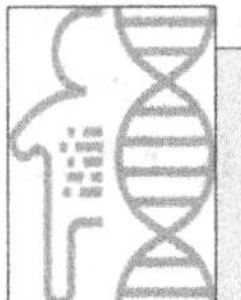

Pharmacogenetic Testing:
- Analyzes how genes affect drug response.
- Example: Testing for variations in the CYP2D6 gene to determine drug metabolism rates for cancer treatment.

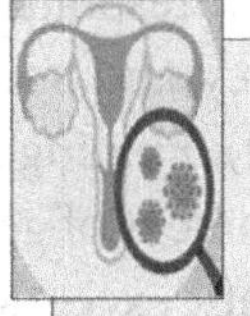

Cancer Genetic Testing:
- Assesses inherited risk factors.
- Example: Screening for mutations in the BRCA1 and BRCA2 genes for ovarian cancer risk assessment.

FIGURE 9.4 Major Types of Genetic Testing.

unknown discoveries that are not related to the original reason for the test. Even if the original goal was to detect genetic explanations for a child's developmental difficulties, these accidental leads might show genetic markers connected to diseases such as cancer propensity or unusual cardiac abnormalities. This vast range of genetic information underlines the complexity and interconnection of our genetic makeup, providing opportunities for proactive health management and personalized treatment that go beyond the immediate focus of the testing.

What makes it challenging despite its importance? The broad acceptance of genetic testing and screening shows potential for improving public health, but it also poses complicated ethical, legal, and societal issues. Balancing individual rights with public health needs is critical. Protecting principles like informed consent, privacy, and anti-discrimination is critical. Attempts to handle genetic data differently, while well-intentioned, may result in injustice and operational issues. While individual ethical rights must be protected, public health ethics advocates for voluntary testing and data sharing for communal health advantages (Tong, 2013). This makes the most sense when considered in relation to reproductive genetic carrier screening (RCS). RCS serves to assess the likelihood of having children with specific genetic conditions, regardless of familial background, before or during pregnancy. The ethics in RCS encompasses several key considerations. Firstly, articulating the goals of RCS initiatives poses a central ethical dilemma. Striking a balance between individual reproductive autonomy and population health objectives is crucial. Critics argue that solely aiming to reduce the incidence of genetic conditions at a population level may inadvertently coerce couples into participating, undermining the voluntary nature of testing. Such an approach may imply an obligation for couples with increased risk to take action, potentially impinging on their autonomy and decision-making. The ethical issues extend to the categorization of gene variations discovered during screening (Dive and Newson, 2021). The ambiguity around the pathogenicity of some variations, especially in the absence of a proband in population screening, creates difficulties. Misclassification can cause stress and needless treatments for couples, but omitting to reveal disease-causing variations might have unintended repercussions for future offspring. The significance of delivering RCS programs as truly voluntary treatments that respect couples' beliefs and choices is highlighted by these ethical issues. As variant databases expand, addressing these ethical difficulties is critical to the appropriate execution of RCS projects. Designing screening panels for such cases requires caution, considering the information's value and feasibility across diverse populations. Regular panel reviews are crucial due to genetic variant complexity (Dive et al., 2022). Similarly, genetic testing for reproductive cancer raises ethical issues, including familial communication responsibilities. Sharing results may burden carriers, posing dilemmas related to privacy, autonomy, and potential familial conflict or discrimination (Ormondroyd et al., 2012). Finding a balance between these viewpoints is critical for navigating the ethical environment of genetic testing.

While we pave the way for genetic exceptionalism, we should also check the appropriate process for obtaining consent for genetic tests, it is necessary to confront the broader question of whether the consequences of the results of those tests are substantively different from the consequences of other "medical" tests, for which specific consent is not always obtained. Some ethicists argue against what has been called the "exceptionalism" of genetic tests. We, as scientists, always believe in progress and advancement, genetic testing will have far greater benefits, and less harm, when done in conjunction with well-designed health education and enlivening experiences, and above all with ethics for genetic testing in place. Long hail healthy genomes.

9.6 FUTURE PROSPECTS

The future of RH and cancer management, alongside other comorbidities, is intertwined with addressing significant challenges. Although genetics, genomes, and environmental factors have all been studied in the past, NGS's full potential has not been realized. These disorders frequently appear in adulthood, thereby suggesting a connection to germline or family abnormalities. There are a lot of unsolved questions as we face intrinsic difficulties. Consequently, it is essential to move toward evolutionary and systems genomics-based methods. There is a critical need to integrate genetic testing with proper administration and education programs.

Genetic testing's widespread acceptance brings potential health improvements but raises complex ethical, legal, and societal issues. Balancing individual rights with public health needs is crucial, protecting principles like informed consent, privacy, and anti-discrimination. RCS highlights these ethical challenges, including coercion concerns and gene variant classification dilemmas. Delivering RCS as voluntary respects couples' autonomy. Ethical considerations persist as variant databases expand, necessitating careful screening panel design and ongoing reviews. Achieving an ethical balance in genetic testing involves addressing these challenges while promoting informed consent and health education. The ethical implications of genetic testing for reproductive organ cancer or mutations extend to its impact on the entire family. Individuals hold the moral obligation of advising family after the results have been disclosed. This "messenger" duty may burden bearers and create discomfort. Effective communication involves several challenges, generating ethical questions about privacy, autonomy, and the possibility of familial conflict or discrimination based on genetic condition.

REFERENCES

Acuin CS, Khor GL, Liabsuetrakul T, et al. (2011) Maternal, neonatal, and child health in southeast Asia: towards greater regional collaboration. *The Lancet* 377:516–525. https://doi.org/10.1016/S0140-6736(10)62049-1

Azimi M, Schmaus K, Greger V, et al. (2016) Carrier screening by next-generation sequencing: health benefits and cost effectiveness. *Molecular Genetics & Genomic Medicine* 4:292–302. https://doi.org/10.1002/mgg3.204

Basoya S, Anjankar A (2022) Cervical cancer: early detection and prevention in reproductive age group. *Cureus* 14:e31312. https://doi.org/10.7759/cureus.31312

Berg JA, Shaver J, Woods NF, Kostas-Polston EA (2022) Women's sexual/reproductive health and access challenges amid COVID-19 pandemic. *Nursing Outlook* 70:238–246. https://doi.org/10.1016/j.outlook.2022.01.003

Böttcher R (2016) Identification of novel prostate cancer biomarkers using high-throughput technologies. Erasmus University Rotterdam. https://repub.eur.nl/pub/94115/161122_Bottcher-Rene.pdf

Capalbo A, Poli M, Riera-Escamilla A, et al. (2021) Preconception genome medicine: current state and future perspectives to improve infertility diagnosis and reproductive and health outcomes based on individual genomic data. *Human Reproduction Update* 27:254–279. https://doi.org/10.1093/humupd/dmaa044

Clarke AJ, Cooper DN, Krawczak M, et al. (2012) 'Sifting the significance from the data'—the impact of high-throughput genomic technologies on human genetics and health care. *Human Genomics* 6:11. https://doi.org/10.1186/1479-7364-6-11

Dawes R, Lek M, Cooper ST (2019) Gene discovery informatics toolkit defines candidate genes for unexplained infertility and prenatal or infantile mortality. *npj Genomic Medicine* 4:8. https://doi.org/10.1038/s41525-019-0081-z

Dive L, Archibald AD, Newson AJ (2022) Ethical considerations in gene selection for reproductive carrier screening. *Human Genetics* 141:1003–1012. https://doi.org/10.1007/s00439-021-02341-9

Dive L, Newson AJ (2021) Ethical issues in reproductive genetic carrier screening. *Medical Journal of Australia* 214:165. https://doi.org/10.5694/mja2.50789

Fasciana T, Capra G, Lipari D, et al. (2022) Sexually transmitted diseases: diagnosis and control. *International Journal of Environmental Research and Public Health* 19:5293. https://doi.org/10.3390/ijerph19095293

Gann PH (2002) Risk factors for prostate cancer. *Reviews in Urology* 4 Suppl 5:S3–S10.

Goetz LH, Schork NJ (2018) Personalized medicine: motivation, challenges, and progress. *Fertility and Sterility* 109:952–963. https://doi.org/10.1016/j.fertnstert.2018.05.006

Gustafson SL, Gettig EA, Watt-Morse M, Krishnamurti L (2007) Health beliefs among African American women regarding genetic testing and counseling for sickle cell disease. *Genetics in Medicine* 9:303–310. https://doi.org/10.1097/GIM.0b013e3180534282

Harper J, Cristina Magli M, Lundin K, et al. (2012) When and how should new technology be introduced into the IVF laboratory? *Human Reproduction* 27:303–313. https://doi.org/10.1093/humrep/der414

Huang H, Cai B (2014) G84E mutation in HOXB13 is firmly associated with prostate cancer risk: a meta-analysis. *Tumor Biology* 35:1177–1182. https://doi.org/10.1007/s13277-013-1157-5

Jiang P, Yue Y (2014) Human papillomavirus oncoproteins and apoptosis (Review). *Experimental and Therapeutic Medicine* 7:3–7. https://doi.org/10.3892/etm.2013.1374

Korenromp EL, Rowley J, Alonso M, et al. (2019) Global burden of maternal and congenital syphilis and associated adverse birth outcomes—estimates for 2016 and progress since 2012. *PLoS ONE* 14:e0211720. https://doi.org/10.1371/journal.pone.0211720

Li H, Ralph P (2019) Local PCA shows how the effect of population structure differs along the genome. *Genetics* 211:289–304. https://doi.org/10.1534/genetics.118.301747

Lindberg LD, Bell DL, Kantor LM (2020) The sexual and reproductive health of adolescents and young adults during the COVID -19 pandemic. *Perspectives on Sexual and Reproductive Health* 52:75–79. https://doi.org/10.1363/psrh.12151

Long KA, Thomas SB, Grubs RE, et al. (2011) Attitudes and beliefs of African-Americans toward genetics, genetic testing, and sickle cell disease education and awareness. *Journal of Genetic Counseling* 20:572–592. https://doi.org/10.1007/s10897-011-9388-3

Mahobia A, Choudhari SG (2022) Social security measures of reproductive health among adolescents in India: a narrative review. *Cureus* 14:e28546. https://doi.org/10.7759/cureus.28546

Mariani A, Seweng A, Ruseng SS, et al. (2021) The relationship between knowledge and personal hygiene and the occurrence of sexually transmitted diseases at the Community Health Center Talise, Palu. *Gaceta Sanitaria* 35:S164–S167. https://doi.org/10.1016/j.gaceta.2021.10.016

Markowitz LE, Dunne EF, Saraiya M, et al. (2024) Human papillomavirus vaccination: recommendations of the Advisory Committee on Immunization Practices (ACIP). *Morbidity and Mortality Weekly Report* 63:1–30. www.cdc.gov/mmwr/preview/mmwrhtml/rr6305a1.htm. Accessed 30 April 2024.

Mayer C, Gasalberti DP, Kumar A (2024) Brachytherapy. In: StatPearls. StatPearls Publishing, Treasure Island (FL).

Mizan GE, Rees D, Wilson K (2016) Reproductive health hazards in laboratory work: back to basics. *Occupational Health Southern Africa* 22:30–33. https://doi.org/10.10520/EJC193376

Ntekim A, Folasire A, Odukoya OA (2023) The prevalence of prostate cancer among young men below 55 years of age in Nigeria. *Cancer Control* 30:10732748231175255. https://doi.org/10.1177/10732748231175255

Oczkowski M, Dziendzikowska K, Pasternak-Winiarska A, et al. (2021) Dietary factors and prostate cancer development, progression, and reduction. *Nutrients* 13:496. https://doi.org/10.3390/nu13020496

Oncology (ESGO) TES of G (2024) ESGO 2017 congress, 4–7 November 2017 in Vienna, Austria. In: *ESGO Gynae-Oncology Guidelines*. https://guidelines.esgo.org/cervical-cancer/guidelines/recommendations/. Accessed 30 Apr 2024.

Ormondroyd E, Donnelly L, Moynihan C, et al. (2012) Attitudes to reproductive genetic testing in women who had a positive BRCA test before having children: a qualitative analysis. *European Journal of Human Genetics* 20:4–10. https://doi.org/10.1038/ejhg.2011.146

Parent C, Tetu C, Barbe C, et al. (2022) Menstrual hygiene products: a practice evaluation. *Journal of Gynecology Obstetrics and Human Reproduction* 51:102261. https://doi.org/10.1016/j.jogoh.2021.102261

Pokhrel D, Bhattarai S, Emgård M, et al. (2021) Acceptability and feasibility of using vaginal menstrual cups among schoolgirls in rural Nepal: a qualitative pilot study. *Reproductive Health* 18:20. https://doi.org/10.1186/s12978-020-01036-0

Poondla N, Madduru D, Duppala SK, et al (2021) Cervical cancer in the era of precision medicine: A perspective from developing countries. *Advances in Cancer Biology - Metastasis* 3:100015. https://doi.org/10.1016/j.adcanc.2021.100015

Ramachandran D, Dörk T (2021) Genomic risk factors for cervical cancer. *Cancers* 13:5137. https://doi.org/10.3390/cancers13205137

Rodrigues VDO, Polisseni F, Pannain GD, Carvalho MAG (2020) Genetics in human reproduction. *JBRA Assisted Reproduction* 24:480–491. https://doi.org/10.5935/1518-0557.20200007

Rowley J, Vander Hoorn S, Korenromp E, et al. (2019) Chlamydia, gonorrhoea, trichomoniasis and syphilis: global prevalence and incidence estimates, 2016. *Bulletin of the World Health Organization* 97:548–562P. https://doi.org/10.2471/BLT.18.228486

Rybak AP, Bristow RG, Kapoor A (2015) Prostate cancer stem cells: deciphering the origins and pathways involved in prostate tumorigenesis and aggression. *Oncotarget* 6:1900–1919. https://doi.org/10.18632/oncotarget.2953

Song C, Kong Y, Huang L, et al. (2020) Big data-driven precision medicine: starting the custom-made era of iatrology. *Biomedicine & Pharmacotherapy* 129:110445. https://doi.org/10.1016/j.biopha.2020.110445

Thoenissen P, Heselich A, Burck I, et al. (2023) The role of magnetic resonance imaging and computed tomography in oral squamous cell carcinoma patients' preoperative staging. *Frontiers in Oncology* 13:972042. https://doi.org/10.3389/fonc.2023.972042

Tong, R. (2012) Ethics, infertility, and public health: Balancing public good and private choice. *The Newsletter on Philosophy and Medicine*, 11:2, pp. 12–17. (Reprinted in Boylan, M. (2014) *Medical Ethics*, second ed., Malden, Massachusetts: Wiley-Blackwell, pp. 13-30.)

Tuddenham S, Hamill MM, Ghanem KG (2022) Diagnosis and treatment of sexually transmitted infections: a review. *JAMA* 327:161. https://doi.org/10.1001/jama.2021.23487

Tvrda E, Agarwal A, Alkuhaimi N (2015) Male reproductive cancers and infertility: a mutual relationship. *International Journal of Molecular Sciences* 16:7230–7260. https://doi.org/10.3390/ijms16047230

United Nations Population Fund (2018) UNFPA – Frequently Asked Questions. https://www.unfpa.org/frequently-asked-questions.

Woodruff TJ, Carlson A, Schwartz JM, Giudice LC (2008) Proceedings of the Summit on Environmental Challenges to Reproductive Health and Fertility: executive summary. *Fertility and Sterility* 89:e1–e20. https://doi.org/10.1016/j.fertnstert.2008.01.065

Yim EK, Park JS (2005) The role of HPV E6 and E7 oncoproteins in HPV-associated cervical carcinogenesis. Cancer Research and Treat 37:319. https://doi.org/10.4143/crt.2005.37.6.319

Index

Printed in the United States
by Baker & Taylor Publisher Services

Printed in the United States
by Baker & Taylor Publisher Services